D. Lichtenstein
General ultrasound in the critically ill

Daniel Lichtenstein

General ultrasound in the critically ill

Forewords by Michael R. Pinsky and François Jardin

With 247 Figures

Daniel Lichtenstein, MD
Hopital Ambroise Paré
Service de Réanimation Médicale
avenue Charles de Gaulle 9
92104 Boulogne (Paris-Ouest)
France
e-mail: dlicht@free.fr

Translation from the French language edition:
L'échographie générale en réanimation by Daniel Lichtenstein
Copyright © 2002 Springer-Verlag France
Springer is a part of Springer Science+Business Media
All Rights Reserved

Library of Congress Control Number: 2007932800

ISBN 978-3-540-73623-3 Springer Berlin Heidelberg New York

This work is subject to copyright. All rights are reserved, whether the whole or part of the material is concerned, specifically the rights of translation, reprinting, reuse of illustrations, recitation, broadcasting, reproduction on microfilm or in any other way, and storage in data banks. Duplication of this publication or parts thereof is permitted only under the provision of the German Copyright Law of September 9, 1965, in its current version, and permission for use must always be obtained from Springer-Verlag. Violations are liable for prosecution under the German Copyright Law.

Springer is a part of Springer Science+Business Media
springer.com
Hardcover edition © Springer-Verlag Berlin Heidelberg 2005
Softcover edition © Springer-Verlag Berlin Heidelberg 2007
Printed in Italy

The use of designations, trademarks, etc. in this publication does not imply, even in the absence of a specific statement, that such names are exempt from the relevant protective laws and regulations and therefore free for general use.

Product liability: The publisher can not guarantee the accuracy of any information about dosage and application contained in this book. In every individual case the user must check such information by consulting the relevant literature.

Editor: Dr. Ute Heilmann, Heidelberg, Germany
Desk editor: Hiltrud Wilbertz, Heidelberg, Germany
Production editor: Andreas Gösling, Heidelberg, Germany

Cover-Design: Frido Steinen-Broo, Pau, Spain
Typesetting: Fotosatz-Service Köhler GmbH, Würzburg, Germany

Printed on acid-free paper 21/3180 ag- 5 4 3 2 1

Foreword

Diagnostic ultrasonography has come of age in the intensive care setting. It has become increasingly common to have training programs teach residents these techniques as part of their fundamental instruction, to have ultrasound equipment available in the ICU and for it to be used by intensivists as an essential tool in the management of the critically ill patient. Ultrasound machines have many applications. They can be used to diagnosis fluid collections in various bodily cavities (e.g., intra-abdominal abscess, ascites, pleural or pericardial effusions) and guide percutaneous catheter insertion. They can be used to diagnose structural etiologies for cardiovascular and respiratory insufficiency, assess intravascular volume status and define cardiac contraction performance. An example of the wide range of structures that can be examined is included in Table 3.2. In fact, considering the non-invasive nature of this form of investigation and the broad number of conditions and uses that ultrasound enjoys, it is surprising that it has taken so much time for this established technique to be embraced by the critical care community.

In this volume, Daniel Lichtenstein has addressed the wide variety of ultrasonic approaches in a rigorous yet easy to read fashion. Ultrasound is a learned technique with numerous specific applications. No volume can be a substitute for hands-on experience and good bedside training. However, as a companion to this training, this book collates in a single volume most of the teaching elements used in ultrasound. The first part of the book addresses in a general fashion the actual technique of acquiring ultrasonographic images of normal biological structures. The second part of this volume focuses on diagnostic aspects of specific organs or bodily compartments. This is the largest section, and nicely separates disease by organ. The section on lung pathology is especially important because it has no counterpart in any other volume on ultrasound. The chapter on cardiac ultrasound is also good but cannot stand alone as the primary information base for echocardiography. However, even though echocardiography is an established sub-specialty of cardiology, this chapter still gives an acceptable overview. The third and final part of the volume addresses the important clinical applications of ultrasound. This part can only describe what the reader must experience first hand, but as a guide, it brings together many important and relevant techniques.

In summary, Daniel Lichtenstein has distilled in one volume a unique and complete description of ultrasound in the critically ill. This singular work by one of the leaders in the field should define the standard from which students develop their understanding and abilities in this important and rapidly growing field.

Michael R. Pinsky, MD, Dr hc
University of Pittsburgh

Foreword

At the beginning of the 1980s, real-time ultrasound, already widely used by cardiologists, arrived in medical intensive care units. At this time, it provided information on cardiac function, and in this domain the use of this method at the bedside provided noteworthy diagnostic short-cuts. In addition, certain physiopathological questions, which often incited interminable discussion, found immediate answers in this new opportunity to directly examine the cardiac chambers.

Contrary to cardiac ultrasound, general ultrasound is not usually part of the working knowledge of intensivists. It is therefore with a certain delay that they realized something obvious: the devices they had acquired to obtain information on the heart were also able to provide much more information in areas that were usually reserved for radiologists. With well-targeted training, the intensivist had a tool that could directly answer the numerous questions raised at the bedside. This book by Daniel Lichtenstein, intensivist and physician-sonographer, perfectly illustrates this progression in the field.

The English edition provides a good deal of new information compared to the first French edition (1992). The author's experience has in fact been enriched by daily work at the bedside of patients hospitalized in Ambroise-Paré Hospital's ICU. Among these new contributions, let us cite in particular lung ultrasound, a discipline that Daniel Lichtenstein has thoroughly described. In addition, a large number of current technical procedures are simplified using ultrasound guidance. In Daniel Lichtenstein's hands, the ultrasound device has become an indispensable tool in the practice of intensive care and emergency medicine.

Prof. François Jardin, PU-PH

Preface

An English translation of *L'Echographie Générale en Réanimation* was necessary, after two French versions in 1992 and 2002.

Ultrasound has, it is true, gained a more important place in emergency and intensive care medicine. Technological evolution alone does not explain this popularity. Technology develops extremely quickly, but we have always suggested – and continue to do so – that before rushing to the most modern ultrasound units, we should already make optimal use of so-called obsolete devices. Since at least 1978, the quality of the images was sufficient to make life-saving diagnoses. One interesting outcome of technological progress is increasing miniaturization, which makes ultrasound easier to exploit in unusual places such as the ambulance or airplane.

Whom is this Book Intended for?

This book has a twofold purpose. Its first aim is to describe the fullest exploitation possible of general ultrasound in the ICU. It is also intended to help popularize a method that remains obscure to those who have never used it. All participants in emergency medical care are therefore concerned.

Junior radiologists called to the ICU or the emergency room to examine a critically ill patient may feel disconcerted, at the beginning, by this type of patient not usually seen in routine practice. In a single volume they will find all and only the information necessary on this very particular patient. The experienced radiologist controls ultrasound. However, our observations have highlighted applications that we did not find in the daily practice of our colleagues, nor in the ultrasound textbooks. The critically ill patient, often on mechanical ventilation, has extremely complex characteristics, resulting in a specific combination of ultrasound signs, and specific interventional procedures following a specific logic, not the logic surrounding the ambulatory patient, a familiar task for the radiologist. Finally, theoretical potentials of ultrasound that have not yet been validated are presented in this book, opening the door to continued research. We ask the experienced reader to be indulgent with the simplifications made in the initial chapters.

On the other hand, intensivists, emergency physicians, and those requesting ultrasound examination wish to penetrate the heretofore impenetrable domain of ultrasound. With ultrasound they will discover wide-ranging possibilities. Furthermore, we must note that ultrasound is increasingly becoming irreplaceable given the desire of intensivists to acquire at least minimal skill aimed at the best possible management of difficult situations, at any time, day or night.

We hope that these pages will provide an informative resource while keeping intensivists informed of the pitfalls resulting from suboptimal use of this tool. This book should in no case be considered as a »pilot's license« but only a didactic aid providing easier access to full use of ultrasound.

The Images

This edition, derived from the first French edition, has collected figures taken with an ADR-4000 (a device that has been available since 1978), in its time facilitating salutary management of countless critical situations, and a Hitachi EUB-405 Sumi (1992). The ADR-4000 and its mechanical sector 3.0-MHz probe, has given images of a lesser quality than more modern devices. We preferred to keep characteristic figures and did not replace these, since a clinically contributive image is definitely better, in the emergency, than the sophisticated image dear to the imaging specialist. The Hitachi 405 with its electronic 5-MHz probe provides latest-generation resolution. All figures in this book have been taken (with a few exceptions indicated) with these devices.

Ultrasound unites elements that cannot be dissociated: the operator, the patient, the machine. The operator's experience is essential. The echogenicity of the patient is also crucial. The quality of the machine comes far behind these two points. The most costly and modern ultrasound device cannot go through bones or airy structures, dressings, nor transform a poorly echoic patient into an echoic one. These are the true limitations of ultrasound. An ultrasound device from the end of the 1970s can save lives.

In the figures, letters are used logically. The letter A indicates any artery, B the urinary bladder, C the colon, D the duodenum, E the stomach, F any fat areas, G the gallbladder, H the heart, I the small intestine, K the kidney, L the liver but also sometimes the lung, M any pathological mass, O the esophagus (*œsophage* in French), P the pancreas, R the rachis, S the spleen, T the trachea, U the uterus, V any vein, and X various organs. The letters RA, RV, PA, LA, LV designate the cardiac chambers (respectively, right auricle, right ventricle, pulmonary artery, left auricle and left ventricle).

These terms are recalled in the figure legends.

In order not to alter the information, the arrows are located at a slight distance (1 or 2 mm) from the indicated target.

One More Point

Before closing this preface, we must note that the present book reflects an experience born of the synthesis of two disciplines that are sometimes distant in their philosophy but inseparable in the daily routine of a hospital: intensive care medicine and imaging. This may explain positions that run against the current of academic teachings: the situation of ultrasound with respect to techniques using ionizing radiations, the importance of maneuverable – but not too small – equipment, the indication of certain interventional procedures, the existence of lung ultrasound signs, etc. We ask readers to be tolerant, since they may feel a discrepancy with what they have been taught.

For instance, applications are described for which radiological signs already exist. Why then complicate things? Let us take the example of pneumothorax. This disorder can, it is true, sometimes be diagnosed with the clinical examination alone, or by radiography alone. Consequently, if considered separately, the ultrasound sign of lung sliding alone may seem anecdotal. Yet if lung sliding is associated with other interdependent signs, progressively building a whole, this whole will take on increasing importance in patient management. Before reaching this stage, an observer of good faith but not fully informed may find certain processes difficult.

In addition, and without evoking overwhelming advantages such as immediate bedside diagnosis at the lowest cost, one issue should be highlighted. The side effects of ionizing radiations are beginning to be better known. Very irradiating techniques such as computerized tomography are very recent (less than 30 years), and the biological and mutagenic effects can just now be understood. It is with a view to a possible future policy of irradiation control that we have desired to make the first move in this direction. Some may judge our position excessive, others will consider it as one of the answers to the principle of precaution.

Finally, this edition is intended to be imperfect and incomplete. The aim of any research is not to provide the absolute and final answer to a question, but rather to try to push back what we could call a twilight zone, in order to decrease the rate of errors. We impatiently await any idea or correction or improvement from those who have made the effort to open this book. These ideas will be welcome and taken into account for the next edition.

Daniel Lichtenstein, MD

Acknowledgements

This book would not have been written without the help of many participants. Nathalie inspired the idea of the first French version. Jean-François Lagoueyte opened the doors to medicine, Bruno Verdière and Gil Roudy held the doors open to the concerns of intensive care. Without them, nothing would have begun.

François Jardin, Yves Menu and Jean-Jacques Rouby were the first to support our project, at a time when this approach to critically ill patients was considered marginal. Let them be warmly thanked.

A circle of faithful colleagues from the first hours of this enterprise took part in most of our studies and have used ultrasound for direct management of their patients. Nathalie Lascols, President of the Cercle des Echographistes d'Urgence et de Réanimation Francophones (ceurf.net), has provided invaluable work. Gilbert Mezière played a first-line role and only his modesty kept his name from the list of authors. Agnès Gepner, Olivier Axler, Philippe Biderman, Christian Mirolo and many others also privileged us with the opportunity to merge scientific work and friendship. Special thanks are due to Alain Koiran.

Others have taken part in or encouraged this project in one way or another, and we would like to thank them: Eve Adeline, Dan Benhamou, Jean Bernard, Parmod Bithal, Gérard Bleichner, Michel Bléry, A. Jay Block, Slah Bouchoucha, Nathalie Boulet, Jean-Marie Bourgeois, Michel Boynard, Uta Braun, François Brivet, Laurent Brochard, Jean Carlet, Pierre Carli, Paul L. Cooperberg, Pierre Coriat, Jean-Paul Courret, Fabrice Czarnecki, Jean-Claude Deslandes, Jean-François Dhainaut, Gilles D'Honneur, Quy Do Dang, Eryk Eisenberg, Konrad J. Falke, François Fraisse, Gérard Friedlander, Guy Frija, Saoul Gayol, Jean-Pierre Goarin, Ivan Goldstein, Patrick Goldstein, Jacques Grellet, Philippe Grenier, Charles Haas, Pierre Haehnel, Laurent Holzapfel, Pierre Kalfon, Roger Kalmanowicz, Renaud Krylatov, Pascal Lacombe, Alan R. Leff, Jean-Jacques Lehot, François Lemaire, Philippe Loirat, William Löwenstein, Isabelle Mabile, Christine Margulis, David Marrache, David Marsh, Jean Marty, Eric Maury, Charles Mayaud, Paul H. Mayo, Henri Mazzola, Georges Mion, Montasser Mouelhi, Linda Northrup, S.A. Pandalai, Paolo Pelosi, Dominique Périès, Jean-Daniel Picard, Michael R. Pinsky, Gabriel Richet, Peter Rinck, Rudolf Ritz, Joëlle Robinet, James A. Russell, Stephen A. Sahn, Cathy Sebat (and the entire team of New Caledonia hospitals), Stanley S. Siegelman, Nicolas Simon, Mohammad Siyam, Michel Slama, Marc Tenoudji-Cohen, Stéphanie van Duin, Philippe Vignon, Jean-Pierre Vilaverde, Jean-Louis Vincent and Francis S. Weill. We again thank the paramedical team of the Hospital Ambroise-Paré who have surrounded this project with affection, and extend our especially warm thanks to the night team.

Finally, we are honored to offer this translated version to Jeannette and Harry, who would not have believed it could be accomplished, and to David and Sidney, who have valiantly lived through the many hours devoted to this exciting but exacting project. Jean-Jacques, this book is finally dedicated to you.

Contents

Part I	**Generalities**	1
1	Basic Notions	3
2	The Ultrasound Equipment	9
3	Specific Notions of Ultrasound in the Critically Ill	13
4	General Ultrasound: Normal Patterns	19

Part II	**Organ by Organ Analysis**	25
5	Peritoneum	27
6	Gastrointestinal Tract	33
7	Liver	41
8	Gallbladder	46
9	Urinary Tract	55
10	The Retroperitonal Space	62
11	Spleen, Adrenals, and Lymph Nodes	66
12	Upper Extremity Central Veins	70
13	Inferior Vena Cava	82
14	Lower Extremity Veins	87
15	Pleural Effusion and Introduction to the Lung Ultrasound Technique	96
16	Pneumothorax and Introduction to Ultrasound Signs in the Lung	105
17	Lung	116
18	Lung Ultrasound Applications	129
19	Mediastinum	134
20	General Ultrasound of the Heart	139
21	Head and Neck	150
22	Soft Tissues	157

Part III	**Clinical Applications of Ultrasound**	161
23	Ultrasound in the Surgical Intensive Care Unit	163
24	Ultrasound in Trauma	165
25	Emergency Ultrasound Outside the Intensive Care Unit	168
26	Interventional Ultrasound	170
27	Emergency Ultrasound and Antibiotic Therapy	175
28	Analytic Study of Frequent and/or Severe Situations	177
29	Learning and Logistics of Emergency Ultrasound	184
30	Ultrasound, a Tool for Clinical Examination	186
31	Concluding Remarks	188
	Glossary	190
	Subject Index	195

**Part I
Generalities**

CHAPTER 1

Basic Notions

Notions of the physical properties of ultrasound are not indispensable for the user. If needed, a reminder will be easy to find in many ultrasound textbooks.

In practice, the mastery of ultrasound follows four steps:

1. Learning to interpret spatial dimensions.
2. The composition of the image: which structures are indicated by gray, dark, and white tones.
3. The descriptive step: this step integrates the first two steps. It consists in the anatomical recognition and the description of the normal structures, then the pathological ones.
4. The interpretative step: this last step depends only on the culture that the operator will acquire, in books or in practice.

Preliminary Note

The ideal material is described in Chap. 2. An ultrasound unit includes several buttons and cursors. The only functions we deem really useful to know at the beginning and for emergency use are:

- The switch-on button (which is not always easy to find)
- The gain setting
- The zoom (widening of the image)

The sole use of these three settings converts the complex device into a simple stethoscope.

Less useful functions in emergency use are, in our experience, the multiple postprocessing choices (we can always use the same one), positive–negative inversions, and annotations. A word must be said on the processing of the image: the best in our opinion is no processing, notably excluding features such as the noise dynamic filter, which is critical for lung analysis. In addition, filters that would make artifacts vanish would also make lung ultrasound impossible. For visual comfort, with increasing experience, smoothing the image can be useful.

The freeze function is an apparently insignificant function. When this button is pressed, it is easy to note that the image previously visible on the screen in real-time immediately looses a large percentage of its definition. This is only a detail, but if one operator provides a static image, and another operator interprets this image (as done in certain institutions), the full potential of emergency ultrasound is not exploited. The philosophy behind our use stems from deactivating this function. Emergency ultrasound is a dynamic discipline, which should be exploited in real-time only. The possibility of freezing the image is useful if measurements are taken.

Step 1: Learning to Interpret Spatial Dimensions

Ultrasound has one particularity: contrary to radiography, CT or MRI, the operator creates the image. This initial weakness progressively becomes a strength. Spatial learning is the first step, the most important, and without doubt the most delicate to acquire, therefore requiring a rigorous approach. The operator must understand how to locate the elements displayed on the screen. The screen is broken down into four parts: upper, lower, left and right (Fig. 1.1).

The upper part of the screen represents the superficial areas. The head of the probe should be imagined at the top of the triangular image. The lower part of the screen represents the deep areas. This is not a source of problems.

Interpretation of the left and the right parts of the screen is more complex, with two interdependent items to control these areas of the screen: a lateral landmark on all well-designed probes and a left–right inversion button. This button should be configured once and for all so that, when the probe is applied to the abdomen, for instance, with the

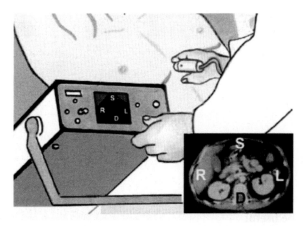

Fig. 1.1. This figure shows spatial configurations. The probe is applied transversally on the abdomen. The right structures of the patient are displayed to the left of the screen (R). The top of the screen corresponds to the superficial area, i.e., the skin (S), the lower part of the screen to deep areas (D). The right diagram shows a reference CT scan

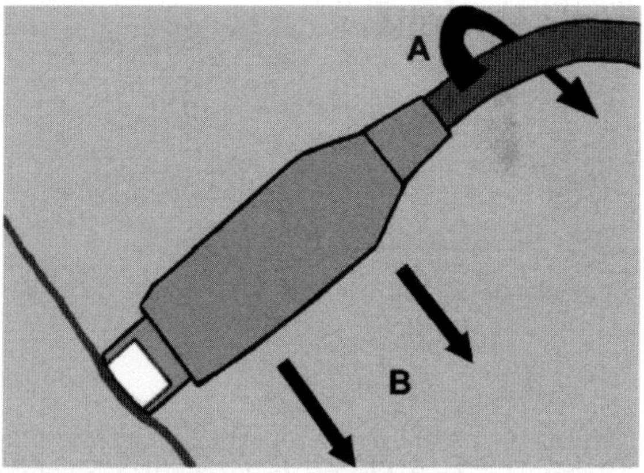

Fig. 1.2. The two main movements of the probe in the space. (A) Rotation along its long axis: the transversal scan becomes longitudinal scan. (B) Scanning: in a transversal approach, the upper then lower structures are displayed

right-sided landmark, the right organs such as the liver should occupy the left side of the screen.

Conventions in medicine serve the purpose of rapid recognition of an image and establishing habits. As a striking example, when a chest radiography is held upside down, the image is extremely unusual and hard to analyze, although nothing has been modified. Therefore, the right organs will always be visualized on the left (of a radiograph, a CT scan, and an ultrasound scan). For transverse scans, the landmark is on the patient's right and the right structures will appear on the left of the screen. For the longitudinal scans, the head should be imagined on the left of the screen, the feet on the right. This is a practical convention in imaging. In cardiology, the opposite convention positions the head on the right of the screen. In our practice, and in order to work uniformly, the head-on-the-left convention is retained, including the heart. In a longitudinal scan of the liver and the kidney, the liver will be seen on the left of the image, the kidney on the right.

An ultrasound scan can be transversal, longitudinal or oblique. At the beginning, the operators should stick to longitudinal or transverse scans. Later, with increasing experience, they will enjoy more flexibility.

Two main movements can be described (Fig. 1.2):

- Scanning, for example, a transverse scan beginning at the epigastrium and ending at the pelvis
- Rotation on its main axis: the study of a vessel on its long axis then on its short axis

These movements create significant changes on the screen, which can be unsettling at the beginning, perhaps the major difficulty of ultrasound, but also its strength. One travels through the so-called third dimension. These changes will be integrated and become automatic with practice. A constructive exercise can be to pass from the short axis to the long axis of a large vessel such as the abdominal aorta.

Step 2: Understanding the Composition of the Image

How does one master the white, gray and black nuances of the images?

Gain

The gain control influences the gray scale. Optimal control of gain is crucial to obtain an interpretable image. Decreased gain will give a black image, and details will be masked. Increased gain will give a white image, and details will be saturated (Fig. 1.3). In the units we use, the proximal, distal and global gains can be adjusted. It should be remembered that only practice provides efficient control of this function. That said, we modify the global gain from time to time, rarely the proximal and never the distal.

A liver–gallbladder scan can be used to establish a proper basis for gray-scale interpretation. The

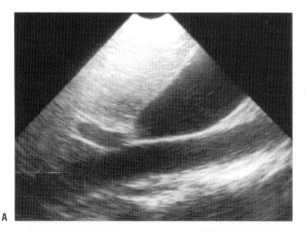
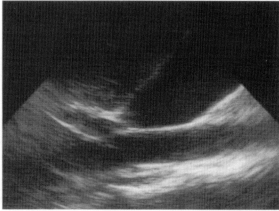
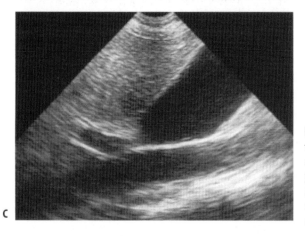

Fig. 1.3. A Longitudinal scan of the liver, the inferior vena cava and the gallbladder. The near gain is too high: superficial areas are saturated. **B** Same scan. The near gain is too low: superficial areas are now underexposed and again escape analysis. **C** Same scan. The gain is optimal. The hepatic parenchyma is now homogeneous, and a good-quality analysis is possible

liver must appear gray, the reference for solid structures. The gallbladder bile must be black, the reference for fluids (Fig. 1.3C), on condition that there is no sludge (see Fig. 8.9, p 50).

Basic Glossary

An anechoic structure yields a black image, since no echo is generated. In the pioneering times of ultrasound, the images were inverted, i.e., anechoic images were represented as white. Logically, increasing the gain does not affect this state.

Inversely, an echoic structure generates echoes, and will give a gray image. It can be more (closer to white) or less (closer to black) echoic.

An image defined as hypoechoic, isoechoic or hyperechoic assumes that a reference image has been defined, usually the liver.

The term »acoustic window« designates a structure that is easily crossed by ultrasound. It thus allows analysis of deeper structures. This window can be physiological (the liver for the study of the kidney or the heart, the bladder for the analysis of the uterus) or pathological (pleural effusion used as an acoustic window to study the thoracic aorta).

The basic term »echoscopy« is used to designate a dynamic sign: cardiac dynamics, ripples, changes in shape, lung sliding, and many others).

Artifacts

An ultrasound image is composed of real and unreal structures. Real structures are anatomical, unreal ones are artifacts. Artifacts are traditionally a hindrance, since they spoil the image. In our opinion, artifacts provide vital information which can be lifesaving. Artifact analysis is, for instance, the basis of lung ultrasonography. Their precise analysis is therefore crucial. Artifacts are created by the principle of propagation of the ultrasound beam. They are stopped by air and bones and accelerated by fluids.

All artifacts converge to the top of the screen, i.e., the head of the probe, like parallels or meridians. They move with the probe movements. They are always regular, straight, i.e., totally different from the majority of the anatomical structures,

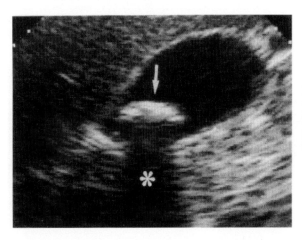

Fig. 1.4. Posterior shadow (*asterisk*) generated by a large calculus (*white arrow*) in the gallbladder. The bile appears anechoic

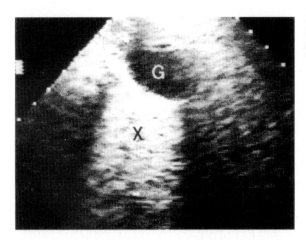

Fig. 1.5. Acoustic enhancement (*X*) arising from the gallbladder (*G*). This sign is of interest when the examined site is poorly defined, since its fluid nature is demonstrated. This is example of a figure that provides the answer to a clinical question (fluid or solid mass?), in spite of the extreme bad quality of the image

which exceptionally are strictly straight and strictly parallel or meridian.

An acoustic shadow is a regular, echo-free (i.e., black) image, which arises from a bony structure (Fig. 1.4). Information behind an acoustic window is thus hidden.

A reverberation echo, repetition echo, or impure shadow is generated by an air structure. A succession of roughly horizontal (or slightly curved) lines generates a striped pattern, alternating dark and clear lines at regular intervals. They can be large (i.e., on the screen, they extend from the left to the right) or very narrow (see Figs. 16.1–16.3 and 16.11, pp 105). A reverberation echo hides the information below, as does an acoustic shadow. At the lung level, acoustic shadows of ribs and reverberation echoes of air regularly alternate, and no information is available under them. If the user is interested not in what happens under the pleural line (a domain that escapes ultrasound) but in what happens at the pleural line, he will realize that artifact analysis can be a discipline in itself.

Acoustic enhancement (Fig. 1.5) creates a more echoic pattern behind a fluid element. For example, the liver parenchyma is more echoic behind the gallbladder than lateral to it. Fluids give acoustic enhancement. In our emergency practice, this familiar artifact is rarely of use.

Last, a small fluid structure such as a blood vessel, surrounded by strongly echoic tissues such as fat will contain various parasite echoes within its lumen (see Chaps. 12–14 devoted to the vessels).

Elementary Anatomical Images

The approach from the anatomical structure (Table 1.1) means:

- A solid tissular mass is echoic: parenchyma, muscle, thrombosis, alveolar consolidation, or tumor.
- A pure fluid mass is anechoic (with acoustic enhancement: circulating blood, vesicular bile, urine, pure fluid collections).
- A pathological fluid mass can be rich in echoes: abscess, hematoma, thick bile, necrosis, etc. If the collection contains tissular debris or bacterial gas, it can be highly heterogeneous. Variations in shape, possible acoustic enhancement,

Table 1.1. Elementary ultrasound images

	Real structure	Artifact
Black tone	Pure fluid	Acoustic shadow
Gray tone	Parenchyma Alveolar consolidation Thick fluid Thrombosis	Acoustic enhancement
White tone	Bone or calculus Fat Valve Interface	Gas

and detection of a dynamic movement within the mass indicate a fluid.
- A gas structure is hyperechoic with posterior echoes of reverberation: air or microbial gas.
- Deep fat is hyperechoic such as mesenteric fat.
- An ossified structure is hyperechoic with posterior shadow: bone or calculus.
- An interface between two anatomical structures results in an hyperechoic strip: pleural layers, diaphragm, cardiac valves, interface between liver and kidney, etc.

The approach from the encountered echostructure means:

- An anechoic image can be:
  - An artifact: the shadow of a bone or calculus
  - A real image: pure fluid
- An echoic image can be:
  - An artifact: acoustic enhancement
  - A real image: normal
    Parenchyma
  - A real image: pathological
    Solid
      Thrombosis
      Alveolar consolidation
      Hematoma
      Necrosis
    Fluid: thickened bile, abscess, noncirculating blood
- A hyperechoic image can be:
  - An artifact: reverberation of gas structure
  - An anatomical structure: surface of a bone, surface of a calculus, surface of a gas bubble, deep fat, cardiac valve, or interface

Step 3: Ultrasound Anatomy: Descriptive Step

Ultrasound anatomy is easy to study when the basic rules of ultrasound and general anatomy are acquired. One difficulty of ultrasound (which also makes it thrilling and unique) is that this anatomy can be studied in the three dimensions. The operator's hand must (or can) make rotating, pivoting or scanning movements, and she thus creates an image to a certain extent. Some clues can be given here. A tubular structure (vessel, bowel loop) will be followed along a certain length when the area is scanned. The practice, at the beginning, of strictly transverse or strictly longitudinal scans will make the images more quickly familiar. With experience, the probe will follow the true axis of the structures. As an example, the analysis of the inferior vena cava can begin in a transverse scan of the epigastrium, probe scanning from the top (head) to the bottom (feet). Once the vein is located, it can be aligned on its long axis by rotating the probe. With experience, it is possible to directly begin by a longitudinal approach to this area, the probe scanning from left to right until the vein is found.

Dimensions can be accurately measured by freezing the image and adjusting electronic landmarks, or calipers. In the figures of this book, the edge of the image displays a centimeter scale, making it possible to measure any structure therein.

How Should One Optimize the Quality of the Image?

Several maneuvers are available and experience plays a major role. Knowledge on how to adjust the gain, for instance, is acquired only with experience. The operator should always attempt to find an optimal acoustic window. For instance, interposition of liver parenchyma between the probe and the heart will optimize, in some cases, subcostal cardiac analysis. In other words, it is sometimes wise to move away from the target to better visualize it. Bones and digestive gas should be avoided. Subcostal organs will be best exposed when the probe is applied strictly against the lower rib.

It is important that the probe remain absolutely still, because the great majority of signs studied in emergency ultrasound is based on dynamic signs. Holding the probe like a pen, with the hypothenar eminence firmly resting on the patient's body, makes this absolute immobility possible and prevents fatigue during a prolonged examination. In addition, this absence of motion will prevent what we call the out-of-plane effect, an effect that creates the illusion of dynamics caused by the movement of the probe alone.

Impediments to Ultrasound Examination

An ultrasound image in a patient with clearly echogenic structures is gratifying, because a definite diagnosis can be made and therapeutic management can be immediately instigated. However, particularly at the beginning, several factors can lead to something of an esoteric fog, which will give this method an unwarranted sense of inaccessibility.

Gas and ribs interrupt the image. This is one of the major drawbacks of ultrasound that is not found with CT and MRI.

Bowel gas is, per se, an inescapable obstacle. However, an acoustic window may exist between two gases. A gas can move, like a cloud hiding the sun. Therefore, before concluding that the examination is impossible, the approaches must be diversified: one must sometimes wait a few minutes and try again. In addition, the operator's free hand may be able to shift the gas.

Air at the lung level is considered an absolute obstacle. We will see that this dogma should be revised.

Bones are absolute obstacles. The adult brain should therefore not be examined with ultrasound. We will see, however, that fine bones (maxillary bones, scapula) are transparent to the ultrasound beam. Using these windows, ultrasound will extend its territory throughout the entire body.

In certain patients, the liver and spleen are entirely hidden by the ribs and cannot be analyzed using the abdominal approach. One must proceed by the intercostal approach, which often results in incomplete images.

Obese patients are not good candidates for ultrasound. This is true for deep structures, where a 3.5-MHz probe will often be inadequate (and a 5-MHz probe even less so). However, paradoxically, critical data can be extracted from analysis of superficial structures. An edifying example is the anterior pleural line, which remains accessible even in extremely obese patients.

A patient covered with extensive dressings will be difficult to examine. This is mainly the case in surgical intensive care units.

In rare young and thin patients, the skin is not a good conductor of ultrasound, and the examination can be disappointing.

In daily practice, an examination that contributes nothing is almost never seen. In fact, ultrasound always answers a clinical question in one of two ways:

1. The item was clearly analyzed, its normal or pathological features are clearly demonstrated, a situation encountered in 80%–90% or more of cases.
2. The conditions are suboptimal, there is a risk of making a mistake, or it is totally impossible to explore a structure, a situation seen in 10%–20% or less of cases.

Step 4: Interpretation of the Image

Only the operator's familiarity with the field, enriched by reading the literature and personal experience (an empirical approach, of interest in a field where much remains to be said), will indicate which conclusions can be drawn from, for instance, a gallbladder wall measured at 9 mm. Previously, this operator has carefully learned to switch on the ultrasound unit, check for the proper gain, locate the gallbladder and take an accurate measurement of its wall.

When no training structure is available, or while waiting for personalized training, two practices can allow one to progress in ultrasound:

- Reading any anatomy textbook will be a good reminder to understand the location of the organs in space.
- Making as many examinations as possible alone and comparing one's conclusions with those of more experienced operators.

CHAPTER 2

The Ultrasound Equipment

The ideal ultrasound device must respond to precise requirements, including simplicity of use, compactness, and mobility. It must be small, equipped with only one probe, and should privilege the two-dimensional.

The Central Unit

The wide range of new equipment available (Doppler, continuous or pulsed, not to mention color and invasive echography) has made the units cumbersome, complex to use, and above all they limit freedom of movement. To justify routine bedside use, the device must be as light as possible but not overly so (see below in this section), arranged on a cart, which is mandatory for easy access to the patient. The idea of having to transport the device should in no case be a physical or psychological obstacle.

The unit should be economically accessible, since the hospital already has sophisticated, high-quality units in the radiology department.

A basic point is that the unit must be compact and smooth, without prominent buttons, or it will not be possible to efficiently disinfect it. The keyboard should have flat pressure keys. A voluminous unit with prominent buttons will certainly be a godsend for germs. The apparatus we have used for several years is highly compact, weights 13 kg, has a 36-dm^3 volume, with advanced esthetics, a flat and easily washable front, with excellent image definition. This technology, which was available before 1994, has contributed to saving countless situations.

Ultraportable »pocket« ultrasound units exist now. One was created many years ago, in a form that we found rather primitive, weighing 3.5 kg and battery-powered [1, 2]. This very rudimentary unit was nevertheless used in the helicopter and was originally, to our knowledge, something of a world premiere [3]. The modern miniature units provide good resolution. In our opinion, these units will make prehospital ultrasound diagnoses easier, i.e., in ambulances, helicopters or airplanes, wherever every square centimeter counts[1]. In the hospital, the »pocket« ultrasound unit has its drawbacks:

1. Its small size makes it an easy prey to theft, potentially a serious problem. If the solution to this problem is to set this miniature unit up on a cart, all the advantages of miniaturization immediately vanish, and only the drawbacks remain (small screen, poor resolution, etc.).
2. In the hospital, we need a minimum of material. The contact product, the disinfection product, the material for emergency interventional procedures, and material for medicolegal purposes (i.e., a hard copy record) are used constantly. Moreover, the device must be set on a stable surface, not be the bed itself. In addition, the larger the screen, the greater the visual comfort. Without all these features, a portable apparatus is not at all practical to use in the hospital, and a cart is, in practice, necessary. The cart cancels the advantage of miniaturization. Without a high-quality image, only the drawbacks remain.

To sum up, in the hospital, a choice must be made between cumbersome and small units, but we believe that choosing between small and ultra-small units would be highly questionable.

---

[1] Note that 40-l ultrasound devices have been available since 1978. This means a 1-cm increase in the length of an ambulance so that the available room inside is not altered. This also means that it was theoretically possible to develop ultrasound in ambulances 30 years ago, long before the development of ultraportable units.

The Probe: Frequency

The frequency of the probe defines its penetration, in the same way that wide-angle or telelenses do this in a camera. The higher the frequency, the better the superficial layers are explored, but the distal ones will be hard to analyze.

Traditionally, a 2.5-MHz probe is used for the heart, a 3.5-MHz probe for the abdominal organs, and a 7.5- or 10-MHz for the superficial organs. In our experience, a 5-MHz microconvex probe alone, covering 17 cm of depth, is ideal for saving lives or – at least – making emergency diagnoses, for the following reasons:

- Experience shows that this frequency opens the door to versatility: see Chaps. 5–30, exclusively illustrated with this frequency.
- All regions of interest in the critically ill patient have one crucial point in common: all these vital organs are superficial. The lung surface is a typical illustration, but also all deep veins, the peritoneum, the gallbladder, the maxillary sinuses, the optic nerves, etc. The heart is a last example, since it is more important to know the status of the ventricles than the auricles.
- From an infection point of view, it seems impossible, or extremely difficult, to carry out an ultrasound examination using several probes without violating asepsis. This basic point is detailed in Chap. 3.
- A complete set of probes is expensive, another critical issue.

Note that in plethoric patients, deep abdominal analysis is often disappointing (the pancreas is a good example), and these patients are eventually referred to CT. Our choice of a single 5-MHz probe has taken this extremely important detail into consideration.

The Probe: Shape

Large surface probes and linear probes should be avoided. Large probes are useful for measuring large organs such as the liver, yet these data are of very little use in emergency settings. A small, microconvex probe seems mandatory in our use (Fig. 2.1). Abdominal organs as well as intercostal spaces (i.e., lung and heart access), complex areas such as the neck, subclavian zone, popliteal zone, etc. can thus be successfully analyzed. Providing an ultrasound unit with only one probe that does not

Fig. 2.1. This probe is light, short, has a 5-MHz frequency, a small surface, and its microconvex shape provides a contact that is curved in one axis and flat in the other. These features are ideal for a versatile use in emergency conditions

provide access to the heart is unthinkable. Last, a small probe is extremely useful for the compression maneuvers and interventional procedures.

The length should be as short as possible. This basic point is important when posterior structures are analyzed: lung and pleura, popliteal veins, etc.

The Issues With Doppler Ultrasound

Doppler ultrasound is a major technological advancement. In our use, however, it is a source of negative side effects. We have considered seven main drawbacks:

1. Learning to use the Doppler ultrasound is long, and gives this modality the justified reputation of being difficult and highly operator-dependent.
2. The higher cost of Doppler-equipped units is an obstacle to wide dissemination.
3. As Doppler units are more fragile, the cost of maintenance and the rate of breakdowns should be taken into account. Repairing also means temporary loss of the ultrasound's contribution to diagnostic procedures.
4. The weight and volume of the usual units are substantial. This can be a limitation for everyday applications, as well as for extreme emergencies.
5. The quality of the image is decreased if Doppler has been added at the expense of the two-dimensional image. This is the usual choice of many manufacturers. What is not a hindrance in cardiac imaging will be a true concern in the

other domains in which higher resolution is required. Every day we see that complex and costly echocardiography units can only explore the heart, whereas simple, low-cost units can provide whole-body documentation, including the heart in its two-dimensional approach.

6. A possible psychological (therefore not to be neglected) effect is that the complexity of Doppler gives an impression of a sophisticated, esoteric examination, and distances emergency ultrasound from what it should be: a simple visual approach to a critically ill patient.
7. Last, the issue of the Doppler device's harmlessness has recently been raised, and has not yet been properly addressed [4, 5].

In evaluating all these potential drawbacks, it is now important to measure the real advantages of Doppler ultrasound. It provides fine information on flow characteristics. The point is to precisely define what is expected in an emergency situation. Our answer will be summarized with two examples. When the question is to determine whether there is venous thrombosis, Doppler is not useful. In a patient with pulmonary edema, determining whether there is mitral regurgitation is important, but rarely alters immediate management. Our experience, gained without Doppler, clearly indicates that the two-dimensional approach provides enough information in emergency situations.

In fact, our belief is that Doppler has very few indications in extreme emergencies. In some cases, emergency Doppler will be useful: arterial dissection, arterial ischemia, etc. In a medical ICU, these concerns are rare enough to justify reflexion. Why not call, in these cases, an outside well-trained specialist with a dedicated ultrasound device? Or, exceptionally, transport the patient to the radiology department? A possible solution that we sometimes use is to have a simple continuous Doppler probe, a very cheap option, which can be combined with the two-dimensional probe with both hands. Basic information can then be provided with minimal investment.

Having a complete Doppler set-up at one's disposal imposes, for few benefits over the year, putting up with all the above-mentioned drawbacks on a daily basis.

Obviously, a device with a solid, inexpensive, small Doppler set-up with no decrease in two-dimensional quality, without an overly complicated keyboard, and with assurance as to its harmlessness, will be major progress for an ICU.

The Cart

The body of the apparatus has a role which goes beyond the simple exterior shell. Four large wheels make for the easiest transport of the device. A handcart-type system should be avoided. A low center of gravity will provide better stability and prevent accidental tip-overs, which should never occur. A place for reprographics or another such system should be present. The kit for emergency ultrasound procedures must have its place. The coupling product and the bottle of disinfectant must also have dedicated places. These details are not mere details, and the growing number of »pocket devices« should not lead to forgetting that the unit alone is not sufficient.

Recording the Data

The hardcopy is useful in order to preserve documents. In extreme emergency situations, a videocassette recorder has two advantages: a shorter examination, since taking photographs is not necessary, and data that can be read subsequently. A film is more telling than static pictures, and will have an obvious didactic impact.

Coupling System

Since the dawn of ultrasound, gel is part of the arsenal. It traditionally provides a coupling between probe and skin, mandatory in order to have a good image. The gel is a long-standing concern. It is cold and sticky, not to say repellent. It makes a good culture medium. Clean-up is difficult and long. Squeezed by the eager hand, the bottle generates gurgling noises which are always uncalled for in dramatic settings. It is highly possible that this gel, an inseparable element of the very image of ultrasound, unconsciously inhibits its expansion.

Several years ago, after an exciting research period, we developed a product making gel unnecessary. Totally harmless, it spontaneously evaporates in a few minutes and leaves no marks. Applied to the area to be explored, it provides the same image quality as the classic gel. All the figures of this book have been acquired using oxygen diprotonate, as we have called it. This product has nothing but advantages. It saves time. It can be used warm, which is greatly appreciated by the conscious patient. Since we have begun using it, ultrasound has become a true pleasure. Just wet a compress with some *tapwater* and leave the gel to the attic… This is one example among others (see lung ultrasound) which shows that simplicity is central to the ultrasound philosophy.

The Problem of Incident Light

Emergency ultrasound has another particularity: it is practiced around the clock. Daylight can be a real problem when it bleaches the screen. Manufacturers do not always think of systems to prevent this inconvenience. We must imagine several temporary set-ups adapted to each unit. The most promising seems to be sliding panels. A matt black cylinder applied to the screen at an oblique angle (towards the operator's eye) is another possible solution.

Disinfection and General Care

The basic problem of the disinfection of the probe and of the areas which are touched during the examination is dealt with in Chap. 3.

The ultrasound unit, the probe, and the cable are fragile objects to be respected.

References

1. Denys BG, Uretsky BF, Reddy PS, Ruffner RJ (1992) Fast and accurate evaluation of regional left ventricular wall motion with an ultraportable 2D echo device. Am J Noninvas Cardiol 6:81–83
2. Schwartz KQ and Meltzer RS (1988) Experience rounding with a hand-held two-dimensional cardiac ultrasound device. Am J Cardiol 62:157–158
3. Lichtenstein D and Courret JP (1998) Feasibility of ultrasound in the helicopter. Intensive Care Med 24: 1119
4. Taylor KJW (1987) A prudent approach to Doppler ultrasonography (editorial). Radiology 165:283–284
5. Miller DL (1991) Update on safety of diagnostic ultrasonography. J Clin Ultrasound 19:531–540

CHAPTER 3

Specific Notions of Ultrasound in the Critically Ill

From the ultrasound perspective, the critically ill patient differs from other patients on two levels. First, since most often comatose and immobilized in the supine position, the patient cannot participate in the examination, maintain apnea, etc. Second, this patient depends more than others on optimal initial management.

The intensive care environment includes both limitations and advantages.

Limitations Due to the Patient's Position

An ambulatory patient can easily be positioned in lateral decubitus with inspiratory apnea for studying the liver, or sitting for studying pleural effusions, or again with legs hanging down for venous analysis, etc. The problem of the critically ill patient has rarely been dealt with in the literature. In this ventilated, sedated patient, the supine position must be exploited to its maximum. The ultrasound approach must be adapted to this position. As a consequence, the procedures described in the following pages are not always purely academic. Their sole ambition is to provide answers to critical clinical questions.

Intensivists performing echocardiography have long adapted their technique by an extensive use of the subcostal route, often the only available route.

Limitations Due to the Material

The critically ill patient is surrounded by impressive life-support materials: ventilator, hemodialysis device, pleural drainage kits, and others. The operator must make sufficient room to work comfortably, a mandatory step for the examination to contribute fully to diagnosis. However, this obstacle can already be minimized by adopting a small ultrasound unit.

The barrier is lowered, thoracic electrodes are withdrawn for heart and lung study (or, to save time, the team should place the electrodes on nonstrategic areas such as the shoulders and sternum), the tracheal tube is gently removed in order to free the cervical areas, the elbows are spread from the chest in order to study the lateral areas (lung, spleen, etc.).

Apnea is difficult to obtain because the patient is mechanically ventilated and cannot hold his breath. A nonventilated patient is often dyspneic or encephalopathic. Experience is sometimes required for this examination, but solutions do exist. Lung ultrasound of a dyspneic patient is perfectly feasible (see Chap. 28). With a little experience, it is even possible to follow the respiratory movements by slight pivoting movements of the probe, and the image will remain stable. In ventilated patients, we rarely need absolute immobility. If needed, lowering the respiratory cycles to a minimum (or simply disconnecting the tube) will be fully effective.

Other Limitations

A sedated patient cannot show pain. Thus, this basic clinical sign will be absent when, for instance, cholecystitis is suspected. One could envisage interrupting the sedation for the duration of the examination, but this procedure remains extremely theoretical. It is better to approach the patient with a different attitude: ultrasound patterns must speak for the patient. This step requires experience as well as a good clinical sense, especially true for the gallbladder.

The Advantages of Ultrasound in a Critically Ill Patient

The critically ill patient is – in a way – a privileged patient with respect to ultrasound. This patient is already sedated, and all interventional procedures will be facilitated. Other remarkable features should render an ultrasound examination optimal.

For instance, mechanical ventilation can allow exploration of organs that were previously hidden, such as an inaccessible gallbladder. When the subcostal approach is limited, it is possible to lower the diaphragm by increasing the tidal volume during a few cycles, as long as there is no risk of barotrauma.

Prolonged intensive care with parenteral feeding can result in a progressive decrease in digestive gas, a feature which considerably increases ultrasound performance.

A hydric surcharge is frequent in septic patients with impaired capillary permeability. This is not an obstacle, since water is a good ultrasound conductor.

The feasibility of ultrasound varies with the patient and with the area. Some examinations are feasible, others are not. Among the feasible ones, some fully answer the clinical question, others do not. In our institution, a study showed a 92% feasibility, all areas combined [1]. The examination was classified as difficult in 29% of feasible cases. The pancreas and abdominal aorta were the main organs that could not be visualized, mainly because of gas (Table 3.1). We should immediately note that in some instances, conditions reputed to make an examination unfeasible can provide precious information. As a simple example, a gas barrier preventing abdominal analysis can indicate pneumoperitoneum perfectly. All in all, one idea should be highlighted: ultrasound examination is always indicated, since only beneficial information can emerge from this policy. To our knowledge, other studies have also found positive data [2].

Developing an efficient system to eradicate bowel gas in any critically ill patient can be of major interest. The idea is to give the physician the best conditions allowing permanent ultrasound access to the abdomen at any time. If such a method exists, is simple and nontoxic, a great step forward will be made in abdominal ultrasonography.

A supine patient offers wide access to the abdomen, the lungs, the majority of the deep veins, the maxillary sinuses, etc. The hidden side of the patient conceals information on very posterior alveolar consolidations, some small pleural effusions, abdominal aorta (lumbar approach), and calf veins. All these areas can, however, be basically assessed without moving the patient, since turning a critically ill patient may sometimes be harmful. To sum up, a whole-body investigation can be performed in the supine position with maximal results.

The critically ill patient has benefited from general ultrasound studies that concluded that this examination was useful [3, 4]. Our hospital is in all

Table 3.1. Feasibility of general emergency ultrasound (expressed as a percentage of cases where the item was analyzed)

Organ	Explorable organ	Optimal exploration	Exploration with a risk of error
Liver	96	72	22
Gallbladder	97	82	14
Right kidney	97	87	10
Left kidney	100	63	37
Spleen	98	75	22
Left pleura (via the abdominal route)	86	54	32
Right pleura (via the abdominal route)	71	58	13
Pancreas	70	51	19
Abdominal aorta	84	51	–
Peritoneum	98	NA	NA
Femoral veins	98	NA	NA
Internal jugular veins	98	95	3
Subclavian veins	93	87	6
Maxillary sinuses	100	100	–
Anterior lung surface	98	98	–
Lateral lung surface	92	86	6
Optic nerve	100	94	6

NA not available.

probability the first to study assessing the usefulness of general ultrasound, handled by the intensivist himself, using a set-up belonging to the ICU [5]. This study showed that, as regards the classic indications of general ultrasound alone, 22% of the patients benefited from a systematic study, with a direct change in the immediate therapeutic management. This study did not take into account negative results (with positive outcome on patient management), cardiac results, interventional procedures and, above all, nonclassic indications such as lungs or maxillary sinuses. If it had, the percentage of patients benefiting from the ultrasound approach would not have been 22% but a number not far from 100%. As an example, the following pages will show that it is possible to decrease irradiation in *every* patient admitted to an ICU.

Conducting an Ultrasound Examination

The region of clinical interest will be studied, of course, but it is often advisable to make a complete examination. The potentials of this noninvasive test are thus exploited as fully as possible. Table 3.2 is a suggestion of an ultrasound report made with this in mind.

In good conditions, in emergency situations, the abdomen, thorax and deep veins can be analyzed in less than 10 min, or even less than 5 min with experience. The examination can be recorded in real-time without losing time taking down figures. A precise answer to a clinical question such as checking for absence of pneumothorax, absence of thrombosis of the vein to be catheterized, absence or presence of bladder distension, etc., usually require only a few seconds.

Table 3.2. Usual ultrasound report

Ambroise-Paré Hospital   Medical intensive care unit
General ultrasound
Name   Date and hour
Operator   Unit: Hitachi 405 Sumi, 5-MHz probe   Age, history
Clinical question(s)
Conditions, echogenicity of the patient
Position of the patient: supine   half-sitting   chair   other
Ventilatory status: spontaneous or mechanical ventilation   PEP   02 level   Eupnea or dyspnea   sedation
Thorax
   Lungs
      Anterior analysis (level 1)
         lung sliding
         air artifacts
      Lateral analysis up to bed level (level 2)
         Type of artifacts
         Pleural effusion
         Alveolar consolidation
      Lateral analysis with posterior extension (level 3)
      Anteroposterior examination in lateral decubitus and apex analysis (level 4):
      Hemidiaphragm: location, dynamics (mm)
   Heart (general two-dimensional analysis)
      Easiness:
      Pericardium: subnormal or not
      Left ventricle
         Diastolic diameter
         Systolic diameter
            Global contractility: impaired   low   normal   exaggerated
            Dilatation: absent   mild   substantial
            Wall thickness
            Other (asymmetry, etc.)
      Right ventricle
         Dilated or not
         Free wall thin or not
         Contractility
      Thoracic aorta (initial, arcus, descending aorta)
      Right pulmonary artery: exposed or not

Table 3.2 (continued)

Abdomen
    Easiness and difficulties (and reasons)
    Fluid peritoneal effusion: absent or other
    Pneumoperitoneum: absent (present peritoneal sliding and/or splanchnogram) or else
    Stomach: full   empty   Gastric tube location
    Small bowel: peristalsis present or abolished or not accessible. Wall thickness. Caliper (mm) normal or distended. Content (echoic or anechoic)
    Colon: air-fluid level search
    Aorta: regular or else
    Inferior vena cava: expiratory caliper at the left renal vein
    Search for thrombosis
    Gallbladder: Painful or not. Global dimensions. Wall thickness. Content (anechoic or sludge or calculi). Perivesicular effusion. Other items (wall abnormalities) or absence of these features
    Liver: no detectable abnormality, no portal gas in partial or exhaustive analysis, or other
    Intrahepatic and common bile duct: caliper (mm)
    Spleen: homogeneous or not, measurement (mm)
    Portal veins: without anomaly or other
    Pancreas: normal size and echostructure or other
    Kidneys: nondilated cavities or other
    Bladder: full   empty
    Uterus
Other
Deep venous trunks
    Two-dimensional ultrasound, without Doppler, with the technique of soft compression
        Internal jugular axes (right or left dominant)
        Subclavian axes
        Inferior vena cava
        Iliac axes
        Femoropopliteal veins
        Calf veins: compressible at least partly (%) or other
Head
    Optic nerves: caliper
        Distal bulge
    Maxillary sinuses (head supine or upright)
        No signal or presence of sinusogram
        If sinusogram present: complete or incomplete
Miscellaneous
    Muscle/adipose tissue ratio
    Others

In practice, a practical synthesis is made from these previously detailed data, with immediate management changes to be envisaged. The clearest answer possible must be given to the clinical question. Positive as well as negative items should be specified. Serendipitous items with immediate consequences must be recalled here.

The style of this report has been adapted for the needs of the book. It contains a maximum of information which will be printed in a minimum of time (a precious advantage in emergency situations). Some data serve as an initial reference for later examinations. The item »normal or other« has been created in order to avoid any ambiguity, if for instance the item was not analyzed.

Disinfection of the Device

Prevention of cross-infections is a major concern in the ICU. It is therefore mandatory to return the ultrasound set-up free of any harmful microbes after examination of a septic patient (or of any patient). One could logically compare an ultrasound examination with a central venous catheterization. In fact, asepsis in ultrasound is not only mandatory, but is above all very easy to follow. A small intellectual investment is the only requirement. We must create automatic reflexes based on logic.

The first step is to define septic areas: the probe and the cable, the keyboard, the lower part of the contact product bottle. These areas will be distinct from the clean areas: the cart, the disinfectant product, and the upper part of the contact product bottle. Septic and nonseptic areas should not be confused during an examination.

With a device set up before any contact with the patient, if the operator touches, after patient contact, only the probe, the keyboard and the contact product, only these three elements will have to be wiped down after the examination (and after hand-washing). A thousand useless gestures should be avoided, such as nonchalantly putting soiled hands on nonseptic parts of the device. The contact product bottle should never lie over the bed; nor should the disinfectant be held by soiled hands. If the device has to be moved slightly, one will use the elbows, or even a foot, but not soiled hands, unless this area is carefully disinfected after the examination, which would be unnecessarily time-consuming.

The disinfectant must remain in a separate place on the cart, for instance on the lower level. It should never be held with contaminated hands. The hands must be washed after having recovered the patient, etc. The probe is then cleaned. It should never be inserted onto its stand before being cleaned. Our procedure is to leave the probe on the bed at the end of the examination, wash our hands, then handle the cable by the end (toward the device) and clean it up to the probe itself. Note that the use of more than one probe will cause serious problems, since a contaminating gesture will be unavoidable. All these steps, at all times necessary, should become automatic and be executed in a precise order. Loss of time will be minimal and the device will remain microorganism-free. The only problem will occur in case of multiple operators: each operator must trust the previous operator. We are fully convinced that, more than 150 years after Semmelweis's first observations, physicians are aware of and take this concern to heart.

Which product should be used? The problem is that the probe must tolerate the product without being damaged. We noted that manufacturers generally provided a vague answer and we have built up experience with the micro-convex, silicone-covered probe of our Hitachi Sumi unit, and a 60° alcohol-based alkylamine bactericide spray with neutral tensioactive amphoteric pH. We have used this system since 1995, and our probe has not shown the slightest damage. Some authors have proposed 70° alcohol as a simple and efficient procedure [6], but a majority of authors find alcohol risky for the probes and not effective enough in terms of decontamination. An aldehyde-based and alcohol-based spray has been advocated [7], but this is a questionable approach since this blend fixes the proteins. Some authors again find that withdrawing all marks of gel with an absorbent towel between two patients is a good solution [8]. In hospital atmosphere and particularly in the ICU, this solution seems highly questionable. All in all, we should not forget that very precise procedures have been established for material disinfection, but since the major problem is removing the gel (a genuine culture medium for bacteria), if gel is no longer used (see Chap. 2), these complicated and constraining procedures should be, to our opinion, forgotten.

Indications for an Ultrasound Examination

When we see the possible benefits of ultrasound for the patient as well as the drawbacks, extensive use of ultrasound clearly is beneficial. Simply admission to an ICU, regardless of the initial possible diagnosis, is an obvious sign of gravity, and justifies a routine ultrasound examination. The question of whether to perform an ultrasound examination should never be raised in a critically ill patient. Too often we have seen cases evolving unfavorably, where the use of ultrasound was late, although it then immediately clarified so-called difficult diagnoses – but too late.

In practice, in our institution, a whole-body ultrasound approach is taken for each new patient admitted to the ICU. It is repeated as many times as necessary. Schematically, three steps can be described:

- The initial step is the initial diagnosis at admission.
- The second step is material management. Interventional ultrasound is of prime importance here (puncture of suspect areas, insertion of catheters, etc.).
- The third step is the follow-up of long-stay critically ill patients, where complications occur (infections, thromboses, etc.).

In each of these steps, ultrasound will play a determining role.

References

1. Lichtenstein D, Biderman P, Chironi G, Elbaz N, Fellahi JL, Gepner A, Mezière G, Page B, Texereau J, Valtier B (1996) Faisabilité de l'échographie générale d'urgence en réanimation. Réan Urg 5:788
2. Schunk K, Pohan D, Schild H (1992) The clinical relevance of sonography in intensive care units. Aktuelle Radio 2:309–314
3. Slasky BS, Auerbach D, Skolnick ML (1983) Value of portable real-time ultrasound in the intensive care unit. Crit Care Med 11:160–164
4. Harris RD, Simeone JF, Mueller PR, Butch RJ (1985) Portable ultrasound examinations in intensive care units. J Ultrasound Med 4:463–465
5. Lichtenstein D, Axler O (1993) Intensive use of general ultrasound in the intensive care unit, a prospective study of 150 consecutive patients. Intensive Care Med 19:353–355
6. O'Doherty AJ, Murphy PG, Curran RA (1989) Risk of Staphylococcus aureus transmission during ultrasound investigation. J Ultrasound Med 8:619–621
7. Pouillard F, Vilgrain V, Sinègre M, Zins M, Bruneau B, Menu Y (1995) Peut-on simplifier le nettoyage et la désinfection des sondes d'échographie? J Radiol 76:4:217–218
8. Muradali D, Gold WL, Phillips A, Wilson S (1995) Can ultrasound probes and coupling gel be a source of nosocomial infection in patients undergoing sonography? Am J Rœntgenol 164:1521–1524

CHAPTER 4

General Ultrasound: Normal Patterns

The term »general ultrasound« is usually understood as abdominal ultrasound. We will accept this rather simplistic view for the time being and provide the physician with a better understanding of the abdominal examination, which can, if necessary, make the very young operator quickly operational. This chapter is highly simplified, including only notions useful in emergency situations.

The abdomen is in fact a modest part of general ultrasound examination, and the reader will find the rest of the body (thorax, veins, head and neck, etc.) described in separate chapters.

It is opportune to adopt a precise order. A possible plan is suggested in Chap. 3, Table 3.2, p 15.

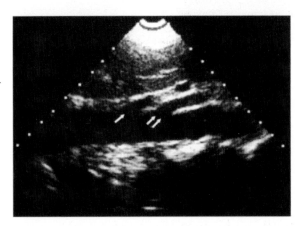

Fig. 4.1. Abdominal aorta, longitudinal view, with the origin of the celiac axis (*arrow*) and the superior mesenteric artery (*arrows*)

Peritoneum

The peritoneal cavity is normally virtual, thus not visible using ultrasound.

Abdominal Aorta

The abdominal aorta descends anterior to the rachis and at the left of the inferior vena cava. Its caliper is regular. The celiac axis and the superior mesenteric artery arise from its anterior aspect (Fig. 4.1).

Inferior Vena Cava

The inferior vena cava rises anterior to the rachis and at the right of the aorta. It passes posterior to the liver (Fig. 4.2) and ends at the right auricle (Fig. 4.3). It receives the renal veins and the three hepatic veins, just before it opens into the right auricle. The walls are rarely parallel, and wide movements are often observed. With all these features, the aorta and inferior vena cava cannot be confused.

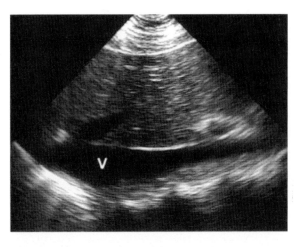

Fig. 4.2. Inferior vena cava (*V*), longitudinal view. Note the bulge (at the *V*), a variation of normal. A measurement of the venous caliper should not be taken at this level

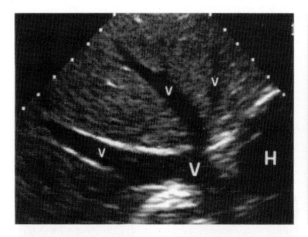

Fig. 4.3. Oblique scan of the liver through the axis of the three hepatic veins (*v*). They meet in the inferior vena cava (*V*), a little before it opens into the right auricle (*H*). Although reputed as having no visible wall, they can, like the right vein here, be separated from the liver by a thin echoic stripe

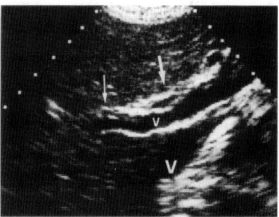

Fig. 4.5. Long-axis scan of the portal vein. The common bile duct (*thick arrow*) and the hepatic artery (*thin arrow*) run anterior to the portal vein. The inferior vena cava (*V*) passes posterior to it

Liver

The liver is studied by longitudinal and transversal scans. Its anatomy is complex to describe, with a right lobe occupying the right hypochondrium, and a smaller left lobe extending to the epigastrium. Radiologists use precise reference scans.

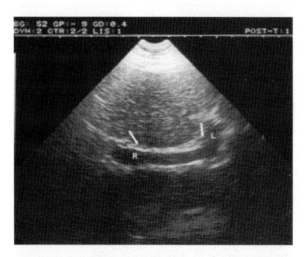

Fig. 4.4. Portal branching, subtransverse scan (slightly oblique to the top and left). This scan shows the right branch (*R*) pointing to the *right*, and the left branch (*L*), also pointing to the *right*. The walls of the veins are thick and hyperechoic, a sign which, among others, distinguishes portal from hepatic veins. Intrahepatic bile ducts are anterior to the portal branching and are normally hardly visible (*arrows*)

Analysis of the hepatic segmentation is complex and finally of little use to the intensivist.

Several vessels cross the liver. Using more or less transverse scans, and from top to bottom, one recognizes:

- The three hepatic veins, which converge toward the inferior vena cava (Fig. 4.3).
- The branching of the portal vein (Fig. 4.4).
- The portal vein, which has reached the inferior aspect of the liver, in an oblique ascending right route (Fig. 4.5).
- The biliary intrahepatic ducts should be looked for just anterior and parallel to the branching of the portal vein (Fig. 4.4).
- The common bile duct passes anterior to the portal vein. Its normal caliper is less than 4 mm (7 mm for some) (Fig. 4.5).
- The portal vein comes from the union between the splenic vein, horizontal, coming from the spleen (Fig. 4.6), and the superior mesenteric vein, visible anterior to the aorta (see Fig. 6.14, p 38).

In longitudinal scans, the liver is visible, from right to left, anterior to the right kidney (see Fig. 4.8), the gallbladder (see Fig. 4.7), the inferior vena cava (Fig. 4.2) and the aorta (Fig. 4.1).

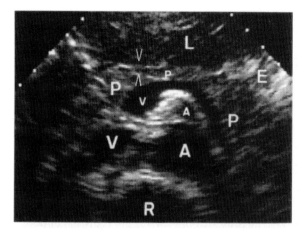

Fig. 4.6. Transverse scan of the pancreas. From rear to front are identified the rachis (*R*), then the aorta (*A*) and inferior vena cava (*V*), then the left renal vein, then the superior mesenteric artery (*a*). Just anterior to it, the splenic vein (*v*) has a comma shape. The splenic vein constitutes the posterior border of the pancreas, which is now located. Its head (*P*) is in contact with the inferior vena cava. The isthmus and body (*p*) are in continuity with the head. Anterior to the pancreas, the virtual omental sac (*arrow*), the stomach (*E*) and the left lobe of the liver (*L*) are outlined. All these structures are rarely all present in a single view

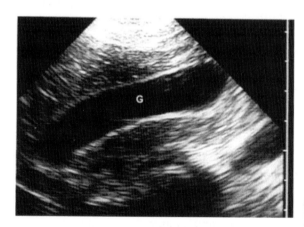

Fig. 4.7. The gallbladder (*G*) usually has a familiar location, at the inferior aspect of the liver, and a familiar piriform shape. It is seen here in the longitudinal axis, has thin walls, anechoic contents and usual dimensions

Gallbladder

The gallbladder is located at the inferior aspect of the right liver, with a piriform shape (Fig. 4.7). It should be sought first in the right hypochondrium, but can sometimes be found in unusual places such as the epigastrium or even the right fossa iliaca. In some instances, it is visible only via the intercostal approach. In order to avoid gross confusions (with a renal cyst, normal duodenum, enlarged inferior vena cava, aortic aneurysm, etc.), one should always locate the gallbladder by first locating the right branch of the portal vein, from which arises a hyperechoic line, the fossa vesicae felleae, which leads to the gallbladder.

Normal dimensions in a normal fasting subject are approximately 50 mm in the long axis and 25 mm in the short axis. The content is anechoic. The wall is at best measured by a transverse scan of the gallbladder. The proximal wall should be preferentially measured. Tangency artifacts should be avoided by making a transversal rather than an oblique scan. A normal gallbladder wall is less than 3 mm thick.

Kidneys

The right kidney is located behind the right liver. From the surface area to the core, a gray then white then black pattern can be described. The gray, echoic peripheral pattern corresponds to the parenchyma. It can vary from average gray (cortex) to darker gray (pyramids or medulla). The white, hyperechoic central pattern corresponds to the central zone, an area rich in fat and interfaces. The dark zone, at the core, is inconstant and corresponds to the renal pelvis, which is normally barely or not visible (Fig. 4.8).

Just under the spleen (Fig. 4.9), the left kidney is less easy to access than the right. It is, however, rare

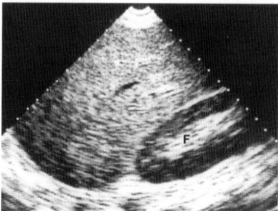

Fig. 4.8. Longitudinal scan of the liver through the right kidney axis. The kidney has a normal size, regular boundaries, a mildly echoic peripheral area, and an echoic internal area (*F*)

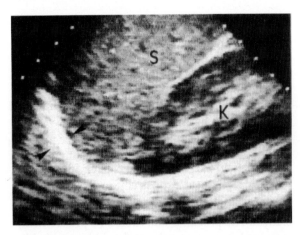

Fig. 4.9. Spleen (*S*) and left kidney (*K*) in a longitudinal scan. Note the left hemidiaphragm (*arrows*) just over the spleen. The kidney is located in the splenic concavity

that no information on the left renal pelvis can be provided.

Over each kidney, the adrenal is normally not identified within the fat (see Fig. 11.9, p 68). Below, the psoas muscle is recognized, with a striated pattern. It descends, vertical, from the rachis to the ala ilii.

Bladder

If empty, it cannot be detected. If half-full, it shows a medial fluid image over the pubic area, with a square section in the transverse scan (Fig. 4.10)

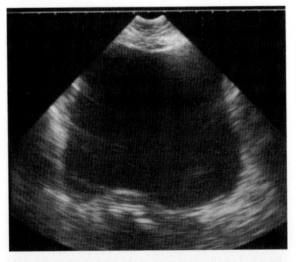

Fig. 4.10. Normal bladder, transverse scan over the pubis. It has a roughly square shape (in fact slightly concave), which indicates moderate repletion

and piriform in the longitudinal scan. When full, the bladder becomes enlarged and round.

Pancreas and Plexus Celiacus

The pancreas and plexus celiacus area is one of the most intricate to master. The surrounding vessels make it possible to recognize the pancreas, with, from rear to front, in a transverse scan, the following ten structures (Fig. 4.6):

1. The rachis, echoic arc concave backward.
2. The inferior vena cava to the right, the abdominal aorta to the left.
3. The left renal vein, oriented horizontally between the aorta and the superior mesenteric artery.
4. The superior mesenteric artery, vertical and thus seen in cross-section. It is easily located since it is surrounded by hyperechoic fat.
5. The splenic vein, horizontal and comma-shaped with a large end to the right, where it receives the superior mesenteric vein and gives rise to the portal vein.
6. The pancreatic gland is then recognized anterior to the splenic vein. The head is anterior to the inferior vena cava. The isthmus and the body are parallel to the splenic vein.
7. The main pancreatic duct can be observed within the gland, horizontal.
8. The virtual omental sac anterior to the pancreas.
9. The horizontal portion of the stomach even farther anterior.
10. The left liver.

The celiac axis is located in a superior plane, and gives the splenic artery to the left and a hepatic artery to the right, which converges toward the portal vein and is applied anterior to it.

Spleen

Located under the left hemidiaphragm, it has a familiar convex and concave shape and is homogeneous (Fig. 4.9). In a supine patient, the probe should be inserted against the bed since the spleen can be more posterior than lateral.

Diaphragm and Pleura

During an abdominal examination, these structures are classically studied through the liver or spleen. The hemidiaphragm and the joined pleural layers form a thick stripe, hyperechoic, concave downward (Fig. 4.9) and stopping the ultrasound beam beyond. We will see in Chaps. 15–18 that this abdominal route is very limited to study the pleural cavity.

Normal Ultrasound Anatomy in a Patient in Intensive Care

To the previous descriptions, one must add the gastric tube, tracheal tube, urinary probe, and central venous catheters. These materials, and others (e.g., the Blakemore probe) will be studied in the following chapters.

Part II
Organ by Organ Analysis

CHAPTER 5

Peritoneum

Detection of a peritoneal effusion or a pneumoperitoneum is routine in the ICU.

The peritoneum covers the major part of the GI tract, abdominal organs, and the abdominal wall. The peritoneal cavity is normally virtual, thus impossible to visualize using ultrasound.

Positive Diagnosis of Peritoneal Effusion

Ultrasound diagnosis of peritoneal effusion is such a basic point that it embodies the place of ultrasound as a tool for the emergency physician. This approach has contributed to saving numerous lives. It suffices to note that the FAST protocol, which in fact has been used for several decades (simply called ultrasound search for peritoneal effusion) has been popularized by the miniaturization of ultrasound units. We strongly believe that this approach could have been available in ambulances since 1978, if ambulances had been extended by one small centimeter (see Chap. 2).

Peritoneal effusion will give a characteristic pattern, provided its analysis is rigorous. It can be recognized by its usually dark echogenicity, location, shape, and dynamic patterns.

1. Dark echogenicity is an accessory sign. In fact, depending on the etiology, the liquid can be anechoic or frankly echoic (pus, blood).
2. Location. In ventilated patients in the supine position, the effusion will collect in five areas (Fig. 5.1). The diaphragm must be localized in order to avoid any confusion with pleural effusion (see Fig. 5.6). The effusion is searched for:
   - Anterior to the liver. One must explore the last intercostal spaces, where the pattern is characteristic (Fig. 5.2). We immediately emphasize this very upper location, at the intercostal spaces.
   - Surrounding the spleen, with the same comment (Fig. 5.3).
   - At the flanks.
   - In the pelvis (Douglas pouch) (Fig. 5.4, and see Fig. 9.13, p 59).
   - Morrison's pouch. In our experience, clinically relevant effusions located at Morrison's pouch, a familiar area but very rarely visible when isolated in a supine patient, are anecdotal or redundant.
3. The shape is highly characteristic. The limits of the collection are concave outside (Fig. 5.5), since they surround the intraperitoneal struc-

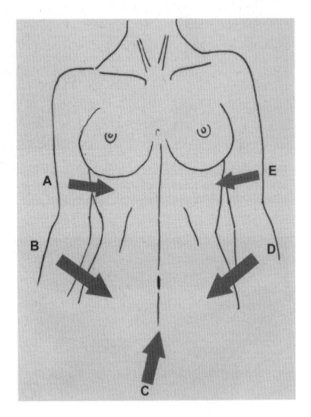

Fig. 5.1. The five areas where peritoneal effusion should be searched for: (*A*) right hypochondrium, (*B*) right flank, (*C*) pelvis, (*D*) left flank, (*E*) left hypochondrium. Note that *arrows A* and *E* are located in the intercostal spaces, not under the ribs as classically done

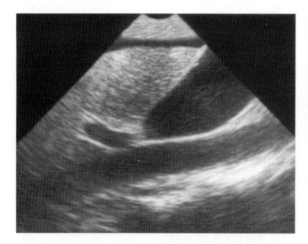

Fig. 5.2. Prehepatic effusion. Here anechoic small effusion, whose thickness varies with the respiratory cycle. A peritoneal effusion can reach this location, and an exploratory puncture at this level is highly contributive

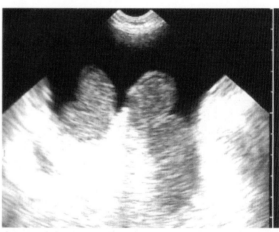

Fig. 5.5. Substantial pelvic effusion. Note its concave limits, which underline the bowel loops. The effusion allows a very fine analysis of the bowel structures. Here, the wall is fine and regular, without villi, i.e., of the ileal type. The content is echoic and homogeneous

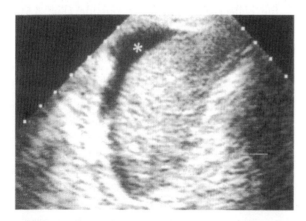

Fig. 5.3. Suprasplenic effusion (*asterisk*). Although minimal, this effusion is clearly identified, with a moon shape between the spleen and diaphragm. Longitudinal scan

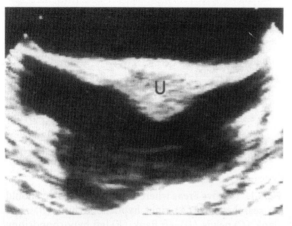

Fig. 5.4. This substantial pelvic effusion isolates the uterus (*U*) and the ligamentum teres. Transversal subpubic scan

tures (liver, gallbladder, urinary bladder, GI tract, etc.). Conversely, encapsulated liquids (gallbladder, urinary bladder, renal cyst, digestive liquid, etc.) have convex limits outside. A dynamic analysis by scanning the area shows that a peritoneal effusion is an open structure, whereas an encapsulated liquid gives a closed shape (this image appears and then disappears during scanning). In practice, a liquid image with concave limits inside cannot correspond to free peritoneal effusion. In the pelvis, a small effusion may simulate, in a hasty test, a half-full urinary bladder (see Fig. 9.13, p 59).

4. Dynamic patterns. A peritoneal effusion can be shaped by the pressure of the probe or by respiratory movements. The bowel loops seem to swim within the effusion.

Ultrasound sensitivity is high for detection of even minimal effusions [1]. A substantial effusion will fill the entire peritoneal cavity and outline the organs. Bowel loops thus become easier to analyze.

Perihepatic effusions can be distinguished from pleural effusions provided the intercostal approach is used, thus first detecting the diaphragm (Fig. 5.6, and see Fig. 15.5, p 98). Nevertheless, if the subcostal route is used, it must be known that only a pleural effusion is located behind the inferior vena cava (see Fig. 15.4, p 97).

Last, ultrasound easily rules out what physical examination can wrongly interpret as an effusion. Ultrasound has often allowed us to avoid inserting a needle in misleading cases, such as a case of

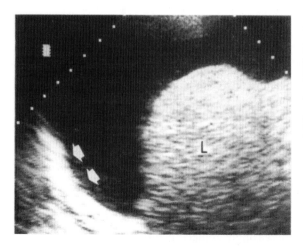

Fig. 5.6. Voluminous suprahepatic effusion, longitudinal scan. The cupola (*arrow*) is separated from the liver (*L*) by the effusion, which means peritoneal location of the effusion. The anechoic pattern of the effusion is suggestive of a transudate

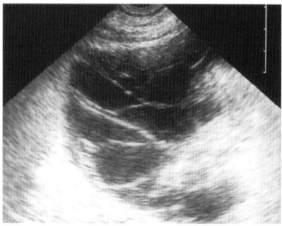

Fig. 5.8. Peritoneal effusion with multiple septations. Patient with peritonitis due to pneumococcus. These septations are rarely visible on CT

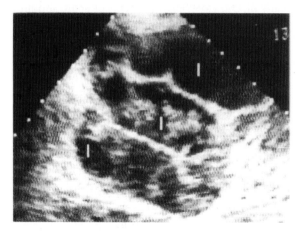

Fig. 5.7. This patient had hydric dullness in the left iliac fossa. An ultrasound examination precluded a puncture, which would have been unproductive or even bloody. It shows absence of peritoneal effusion and several agglomerated bowel loops (*I*) with fluid inside

agglutination of bowel loops with liquid contents, which gave dullness of the flank (Fig. 5.7).

Diagnosis of the Nature of the Effusion

An anechoic effusion generally means transudate, though exudate or hemoperitoneum can produce the same pattern. Anechoic peritoneal effusion is a very frequent finding in an ICU (38% in our series), sometimes limited to a small subphrenic location. This usually occurs when there is an obstacle to venous return (e.g., mechanical ventilation, right heart failure), capillary leakage, or portal hypertension. Most of these etiologies have characteristic ultrasound patterns (right heart dilatation, hepatic structure of cirrhosis, etc.). Peritoneal effusion in a patient suffering from anasarca should not, in principle, be punctured, but we have a more flexible attitude with this (see Interventional Ultrasonography Chap. 26).

Effusion containing a multitude of echoes in suspension, as if in weightlessness, with dynamics in rhythm with respiration cannot be a transudate. Frank hemoperitoneum, peritonitis but also hemorrhagic ascites will give this pattern (see Fig. 5.9). One could call this sign the sign of the internal dynamics, or better yet, the weightlessness sign, but we have retained the plankton sign (see Chap. 15).

Effusion with multiple septations indicates inflammatory effusion, generally, peritonitis (Fig. 5.8). Note that these septations are not visualized with CT.

Although the echostructure of an effusion can guide the diagnosis, our outlook is to practice easy puncture, since the excellent risk-benefit ratio makes this procedure particularly safe (see Interventional Ultrasonography Chap. 26).

Hemoperitoneum

Patterns showing hemoperitoneum can be various. It can produce anechoic collection, can display a multitude of slowly moving echoes as if in weightlessness, or plankton sign, which is immedi-

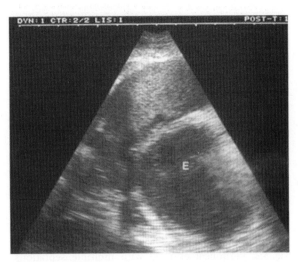

Fig. 5.9. In a longitudinal scan of the left hypochondrium, this mass, which may simulate a spleen in an exclusively static analysis, is in fact moving in a slow rhythm (plankton sign). This pattern is the one of a recent hemoperitoneum. *E*, stomach

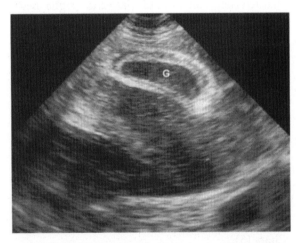

Fig. 5.10. The gallbladder (*G*) of this patient is surrounded by a mass with an apparently solid pattern. This is in fact a clotted hemoperitoneum

ately suggestive (Fig. 5.9), and can also appear as a large echoic, heterogeneous mass, caused by early clotting (Fig. 5.10 and see Fig. 9.19, p 60). In this case, the collection appears solid and one of its main characteristics, variations in shape, is no longer found. It can thus be confounded within the abdominal contents, which melts bowel loops, omentum and various types of fat. Figure 20.24 proves that the blood can alter its echogenicity in a few seconds. In some subtle cases, clotting appears by successive layers, and can give the illusion of bowel loops. This pattern, which can appear early, could be problematic, since abundance and even the existence of the hemoperitoneum can be inadequately assessed. This trap can be easily bypassed, however, with intercostal scans. Observation shows that a majority of cases of clotted hemoperitoneum have a double component, with an upper liquid that usually collects in the very superior areas. A puncture, possibly within the intercostal space, can on occasion confirm the diagnosis.

On some occasions, ultrasound can show the origin of the bleeding: splenic or hepatic rupture, for example.

Note that in the trauma context, ultrasound is increasingly replacing the traditional diagnostic peritoneal lavage [2].

Peritonitis

Perforating peritonitis is a constant risk in the critically ill. Our experience in terms of acute abdominal disorders shows that physical examination, especially in sedated, aged patients, is notoriously insufficient. Bedside plain abdominal radiographs, always hard to obtain, generally generates useless irradiation.

Observation suggests that, in a patient with any acute diagnostic problems, detection of peritoneal effusion is decisive. Minimal effusions are generally more suspect that substantial ones. Secondary development of a peritoneal effusion in a patient whose hydric balance is maintained negative is also suggestive of a complication.

The pattern of the effusion is suggestive when it is echoic or has multiple septations (see Fig. 5.8). Echoic layers surrounding the bowel loops are seen when there is formation of pseudomembranes. Presence of gas within the collection [3] seems a rare observation. Once more, a policy of easy puncture can substantially clarify a clinical situation that was complex or caused hesitation. Surgical decisions can be made before clinical signs become obvious.

Bowel analysis can also be rich in information that can accelerate the decision for surgery (see Chap. 6). Thickened walls and abolished peristalsis are some of the basic anomalies.

Last, detection of pneumoperitoneum will be decisive here (see next section).

Pneumoperitoneum

Ultrasound's potential to detect pneumoperitoneum is rarely exploited. The literature describes an air barrier with a linear shape and acoustic shadow in the extradigestive situation [4], visible under the left liver, surrounding the gallbladder, in the Morrison pouch. However, the abdomen is rich in gas structures, and more precise signs should be described.

1. Gut sliding [5]. It is possible to observe a sliding movement, in rhythm with respiration, at the abdominal level, which obviously corresponds to the two peritoneal layers coming in contact. We called this sign gut sliding in the interest of brevity (Fig. 5.11). Gas collects in the nondependent area of the abdomen, i.e., in highly accessible areas in a supine patient. In a personal study, gut sliding was present in 92 of 100 cases in normal subjects, and it was abolished in all seven confirmed cases of pneumoperitoneum [5].

   These data show that gut sliding can be abolished in various conditions, for example, because of peritoneal symphysis after some surgeries, or because of an abolition of the diaphragmatic course in critically ill patients. A distended stomach will come against the anterior wall, making gut sliding hard to detect. Consequently, analysis of gut sliding will contribute more if the stomach was previously localized in one way or another.

2. Splanchnogram [5]. An extremely contributive sign when gut sliding is absent is the detection of anatomical structures such as the liver or bowel loops (see Figs. 5.2–5.10), a familiar pattern we called the splanchnogram in this context when the probe is applied in a supine patient in a sky-earth direction. It can refer to the liver or even to the mesenteric fat, and can be called a hepatogram or steatogram, for instance. This observation obviously proves that no gas structure is interposed between the abdominal wall and the visceral structures. A gas collection would yield a complete acoustic shadow. In a personal study, absence of splanchnogram predicted pneumoperitoneum with a 100% sensitivity.

3. Other signs. Horizontal lines arising from the peritoneal line are a basic sign of pneumoperitoneum, very sensitive, although not specific. Detection of a peritoneal point is a very specific sign. An equivalent of these signs is described in Chap. 16, devoted to pneumothorax, since the principle is the same.

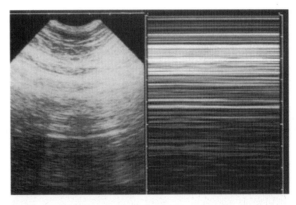

Fig. 5.11. Pneumoperitoneum. *Left* (real-time): massive air barrier. *Right* (time-motion): this mode objectives the complete absence of gut sliding

In acute abdominal disorders, ultrasound can replace the traditional radiograph showing cupolas or the positional radiographs, which are irradiating and tiring (not to say dangerous when the patient is asked to be upright). It is highly logical that ultrasound is more sensitive than radiography for early pneumoperitoneum.

In practice, note that a conserved peak gut sliding or the visualization of visceral structures in a sky-earth approach of the abdomen, allow pneumoperitoneum to be discounted, at the bedside.

Interventional Ultrasonography

When working with peritoneal effusion, we practice ultrasound-assisted puncture at the slightest doubt. Free of complications when done properly, it has an excellent risk-benefit ratio. This is especially true in the critically ill patient, whose clinical data rarely lead to a clear diagnosis. We find this attitude paradoxically safer than the always risky attitude of inferring the type of effusion from its echostructure. In our routine, basic diagnoses are regularly made, in spite of a misleading clinical presentation.

We almost always use a 21-gauge green needle. One major advantage of ultrasound is that one can puncture far from the traditional landmarks. It should be remembered that an intercostal tap can be highly contributive. A tap in the right iliac fossa is classically forbidden, but is for us very commonplace: ultrasound shows that a liquid collection is interposed before the cecum. Ultrasound even makes it possible to puncture the forbidden area located at the level of the epigastric vessels, since these vessels can be clearly identified (Fig. 5.12).

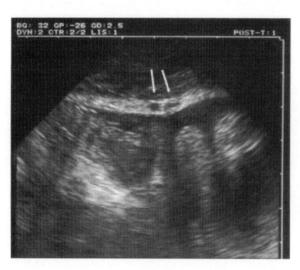

Fig 5.12. Transverse paraumbilical scan. Two tubular parietal structures can be seen: the epigastric vessels (*arrows*). Note the peritoneal effusion deeper

The procedure itself is simple: one almost always performs the tap just after ultrasound location (see Chap. 26).

It is sometimes difficult to puncture very localized effusions in the pelvis of elderly patients. One hypothesis, which seems confirmed by real-time ultrasound observation, is that the needle drives back a loose parietal layer without piercing it. In this case, persisting in inserting the needle to the end could lead to piercing undesirable structures (bowel loops, iliac vessels, etc.). Ultrasound guidance is required but, even here, some procedures remain delicate.

References

1. Ferrucci JT, Vansonnenberg E (1981) Intra-abdominal abscess. JAMA 246:2728–2733
2. Rose JS, Levitt MA, Porter J et al (2001) Does the presence of ultrasound really affect computed tomographic scan use? A prospective randomized trial of ultrasound in trauma. J Trauma 51:545–550
3. Taboury J (1989) Echographie abdominale. Masson, Paris, pp 246–249
4. Gombergh R (1985) Atlas illustré des indications classiques et nouvelles de l'échographie. Polaroïd, Paris
5. Lichtenstein D, Mezière G, Courret JP (2002). Le glissement péritonéal, un signe échographique de pneumopéritoine. Réanimation 11 [Suppl 3]:165

CHAPTER 6

Gastrointestinal Tract

Ultrasound analysis of the GI tract is not routine and is rarely listed in abdominal ultrasound reports. The bowel is, in fact, often considered a hindrance to the analysis of deeper structures. However, its analysis can be decisive in the critically ill. Bowel analysis, it is true, is conditioned by the presence of gas, and is somewhat hazardous (Fig. 6.1). Nevertheless, it is extremely rare that one cannot see at least a small part of the 8 m of the abdominal bowel. Nearly every part of the GI tract can be disturbed by acute disorders.

Normal Ultrasound Anatomy

Bowel wall thickness, practically unchanged from the stomach to the colon, ranges from 2 to 4 mm [1]. Some authors describe several layers [2].

Abdominal Esophagus

The esophagus penetrates the abdominal cavity just anterior to the aorta. The frank acoustic shadow of a gastric tube serves as a practical landmark (Fig. 6.2).

Stomach

The vertical portion, or fundus, passes between the liver and spleen (Fig. 6.3). It is often hard to visualize by the anterior approach and we study it by a lateral, trans-splenic approach. It can be observed in the concavity of the spleen.

The horizontal portion, or antrum, should be investigated by the epigastric approach, with a rounded pattern when empty, or enlarged when the antrum is filled with liquid (Fig. 6.4).

Duodenum

The duodenal bulb follows the pyloric stricture. The second duodenum descends vertically at the contact of the gallbladder and surrounding the

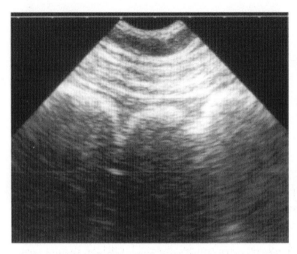

Fig. 6.1. The probe is applied on an abdomen affected with meteorism. No deep structure can be identified, since digestive gas stops the progression of the ultrasound echoes

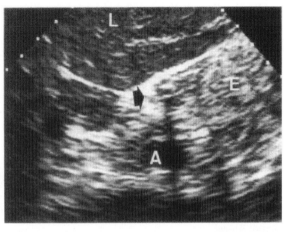

Fig. 6.2. Abdominal esophagus (*arrow*) anterior to the aorta (*A*), behind the left hepatic lobe (*L*) and continuing up to the stomach (*E*). The frank posterior shadow arising from the gastric tube (*arrow*) gives a precise landmark. Transversal epigastric scan

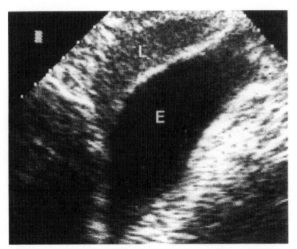

Fig. 6.3. Vertical portion of the stomach (*E*), clearly outlined by an anechoic fluid content. Longitudinal scan. *L*, left hepatic lobe

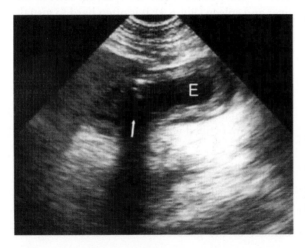

Fig. 6.4. Horizontal portion of the stomach (*E*), just under the liver. One can precisely measure the wall thickness, describe an anechoic fluid content, and localize the gastric tube (more by the frank acoustic shadow [*arrow*] than by the tube itself). Epigastric longitudinal scan

pancreas head. Duodenum patterns are variable and should not be confused with pathological collections. A prolonged observation will show filling and emptying movements. The third duodenum is visible between the aorta and the superior mesenteric artery.

Small Bowel

It is almost always possible to visualize at least some loops of the small bowel. The jejunum is recognized by the endoluminal presence of villi (see Fig. 6.15). The ileum has a tubular, regular pattern (see Fig. 5.5, p 28). Observation shows that acute disorders of the bowel affect the whole of the bowel. Consequently, ultrasound analysis of an even small portion can be rich in information. Many relevant items can be extracted:

1. Peristalsis gives a permanent crawling dynamics, with regular contractions [3]. A present peristalsis can be objectified in a few seconds. This is the usual pattern in the normal subject. Prolonged observation (at least 1 min) seems necessary to affirm abolition of peristalsis.
2. Cross-sectional area. in our observations, the normal caliper of the small bowel is approximately 12–13 mm.
3. Contents can have either a homogeneous echoic (see Fig. 5.5, p 28) or hypoechoic pattern (see Fig. 6.15). The clinical relevance of this distinction is being investigated.
4. Wall thickness ranges from 2 to 4 mm [1]. Fine analysis of the wall is greatly facilitated when there is liquid contrast from both sides, i.e., peritoneal effusion associated with fluid content, two conditions often present in acute disorders (see Fig. 6.15).

Colon

The colon is a tubular structure with visible haustra (Figs. 6.5 and 6.6), without identifiable peristalsis. Roughly, the ascending and descending colon

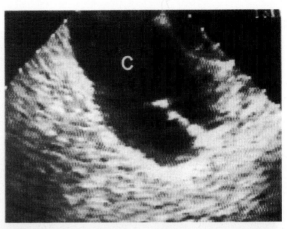

Fig. 6.5. The cecum (*C*) in a longitudinal scan. Fluid sequestration makes the cecum easy to identify. The entire GI tract is filled with huge amounts of fluid in this patient in shock, reflecting major hypovolemia. This disorder should be exploited, since it allows fine analysis of the digestive wall

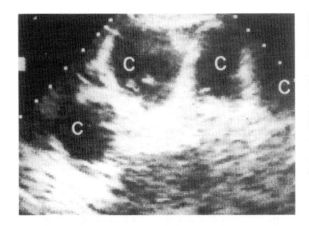

Fig. 6.6. Round, anechoic images, piled up along the left flank in a longitudinal scan (*C*). A slight movement of the probe shows that all these images communicate, and demonstrate this is the descending colon and its haustra

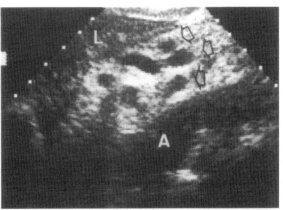

Fig. 6.7. Esophageal varices. In this longitudinal scan, several tubular anechoic images that communicate with each other along the lesser omentum (*arrows*) can be observed behind the liver. These are stomachic coronary varices (*L*, liver; *A*, aorta)

are vertical structures located in the flanks, the transverse colon is horizontal at the epigastric level and distinct from the stomach [4].

The rectum seems, for the time being, without ultrasound interest in emergency medicine.

Abdominal Esophagus

Ultrasound holds a modest place behind fibroscopy. However, esophageal varices are accessible to ultrasound: they give sinuous tubular anechoic structures along the lesser omentum, a hyperechoic area located inside the smaller curvature of the stomach (Fig. 6.7).

With GI tract hemorrhage, detection of esophageal varices cannot be blamed for their rupture and thus the cause of bleeding, but can help in deciding whether major bleeding requires blind life-saving esophageal tamponade.

In addition, ultrasound can provide other signs of portal hypertension (see Chap. 7).

A Blakemore-Linton tube can be inserted with ultrasound guidance. The intragastric position of the tube, before filling, can be detected by visualizing the acoustic shadow, which is frank, tubular and unique. The gastric balloon can then be inflated. It looks like a large, round image, convex outside, highly echoic, with a frank acoustic shadow. The tube is then pulled to the head until resistance is encountered. The gastric balloon becomes visible at the top of the fundus (Fig. 6.8). The esophageal balloon can then be inflated. It will create a mark behind the left auricle (see Fig. 19.10,

p 137). Monitoring thus with ultrasound is quick and very reliable if the operator is trained and the patient has favorable echogenicity.

Stomach and Duodenum

Ultrasound analysis of the stomach can provide a great deal of information. Checking for vacuity or repletion is a first application, which requires only a few seconds in good conditions. For instance, it can be theoretically possible to determine whether

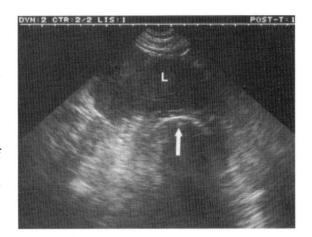

Fig. 6.8. This arciform structure that stops the echoes (*arrow*) is the gastric balloon of a Blakemore tube. On echoscopy, one can see it stumble upward when traction is exerted on the tube, since it outlines the gross tuberosity, the very aim of the procedure. Epigastric transversal scan. *L*, liver

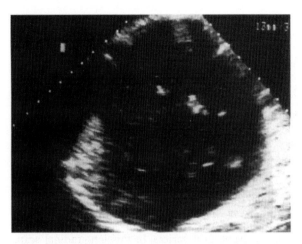

Fig. 6.9. Major fluid stasis with acute gastric dilatation. The content is heterogeneous with hyperechoic points due to alimentary particles. Epigastric transversal scan

this patient can be operated before the traditional 6-h fasting. One can also search for a residue during enteral feeding or diagnose acute gastric dilatation in a patient with acute abdominal disorder. Acute gastric dilatation is a rare but possible cause of acute dyspnea, which gastric aspiration alone can relieve.

Gastric liquid retention gives a massive collection with multiple echoic particles, like in weightlessness, and sometimes an air–fluid level (Fig. 6.9). This pattern is sometimes impressive and can be unsettling for the young operator, and should not lead to diagnoses such as splenic abscess. In our experience, very substantial liquid stasis was often associated with bulbar ulcer, a feature already described in the literature [5].

The correct positioning of a feeding tube within the gastric lumen can be assessed, or alternatively with the mandatory radiograph. Its tubular structure with frank acoustic shadow is easily recognized (Fig. 6.4). This application is very contributive when the end of the tube is at the antrum level, far less when it remains in the fundus area.

Gastric ulcer can produce a thickened, irregular wall. The ulcer itself is rarely highlighted. Ultrasound will not replace fibroscopy, but represents an initial approach that should be validated.

The stomach can be used as an acoustic window for exploring deeper structures such as the pancreas. The stomach should be filled with water, using the gastric tube that is usually present. A slight right decubitus will trap the air bubbles in the vertical portion of the stomach [6]. Last, a full stomach can be precisely located in the still hypothetical aim of performing bedside gastrostomy under sonographic guidance.

A duodenal ulcer will be suspected when a thickened wall is associated with gastric stasis [5]. A study based on 20 cases of duodenal ulcer found an average 7 mm of thickening and reported a sensitivity of 65% and a specificity of 91% for ultrasound [2]. In the case of fluid collection outside the duodenum with gas bubbles, or pneumoperitoneum (see Chap. 5), the diagnosis of complicated ulcer (with leakage) is probable [7].

In caustic intoxications, ultrasound can detect diffuse edema along the GI tract, with a thickened and hypoechoic wall. Search for a left pleural effusion (present if there is esophageal rupture) or peritoneal effusion is part of the initial examination and the follow-up of the patient.

Ultrasound's contribution in GI tract hemorrhage is detailed in Chap. 28.

Small and Large Bowel: Introduction

Here again, ultrasound can play a priority role, when compared to physical examination, plain radiographs, colonoscopy or even CT. In the ICU, a basic contribution of ultrasound is its ability to detect the presence or absence of peristalsis (Fig. 6.10). This information should be considered crucial. Observations have shown a high correlation between abolished peristalsis and the existence of an abdominal drama such as mesenteric infarction or GI tract perforation.

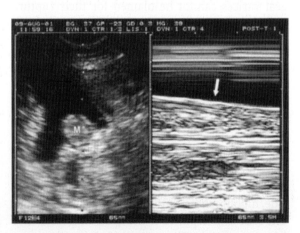

Fig. 6.10. These oblique lines (*arrow*), which seem to intersect in time-motion, are typical from a normal peristalsis. Direct observation in real-time shows the same pattern. *M*, bowel loop surrounded by effusion

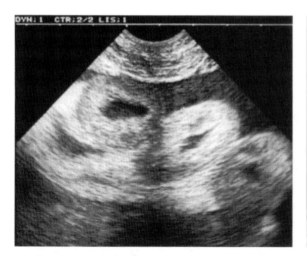

Fig. 6.11. Three bowel loops are visible in cross-section. Note the substantial wall thickening, which can be accurately measured between a peritoneal effusion and anechoic fluid digestive content

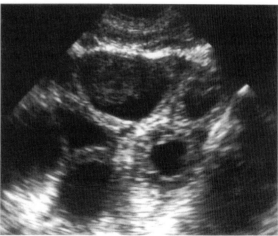

Fig. 6.12. Mesenteric infarction. The entire small bowel has the same pattern, with moderately thickened wall, and above all complete absence of peristalsis. This general pattern of akinesia is striking in real-time. Note the fluid content of the bowel loops. Pelvic scan

Another accessible item is wall thickness measurement (Fig. 6.11). Parietal thickening is present in many critical situations. Doppler could find a place if searching for signs of good perfusion [8, 9], but this is probably of little relevance and may be redundant, at least in the ICU setting.

Small and Large Bowel: Acute Ischemic Disorders

We have grouped different disorders such as mesenteric ischemia, mesenteric infarct or necrosis, colic ischemia and colic necrosis into this single section. The problem lies in the difficulty of the diagnosis, which usually results in delayed treatment and a poor prognosis. Colonoscopy or even CT are not perfect tools. CT can yield troublesome false-negative tests.

In this context, ultrasound deserves a top-ranking place according to our experience. Our observations show a complete and diffuse abolition of peristalsis in 87% of our cases [10]. A moderately thickened wall (5–7 mm) is found in only half our cases (Fig. 6.12). Peritoneal effusion was present in half of cases. Portal gas, a quasi-specific sign, was rarely observed (see Fig. 7.2 and 7.3, p 42).

We must therefore detail the signs demonstrating peristalsis. Observation shows that a patient who is intubated, mechanically ventilated, and sedated with high-dose morphinomimetics, has maintained peristalsis. Adding a curare does not abolish the ultrasound peristalsis. The notion of sedation or even curarization should therefore never be retained to explain an akinetic bowel. The notion of recent laparotomy, even with the procedure touching the bowel, should not be pretext for a wait-and-see policy, since we have observed peristalsis of the small bowel clearly present 24 h after colectomy. Last, for still unknown reasons, a small percentage of ICU patients (12%) without GI tract impairment show abolition of peristalsis.

In the case of colic ischemia, our observations often show thickened colic wall (Fig. 6.13). In addition, small bowel peristalsis is nearly always abolished, a finding that can appear beneficial for an early diagnosis.

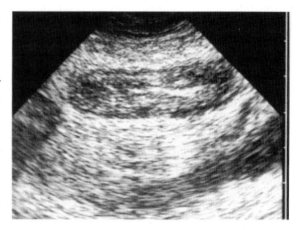

Fig. 6.13. Cross-section of the descending colon. The lumen is virtual, but the wall can be accurately measured, here to 7 mm. Colic ischemia

Chapter 6 Gastrointestinal Tract

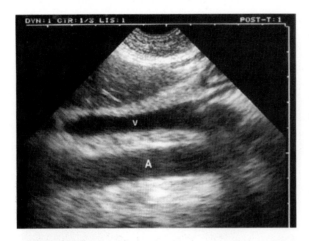

Fig. 6.14. The superior mesenteric vein is often clearly visible (*V*), passing anterior to the abdominal aorta (*A*). The two should not be confused. The good quality of the picture makes it possible to study its content, here anechoic. A local compression maneuver completely collapses the venous lumen. Longitudinal view

The literature is not particularly informative in this field [11, 12]. It describes dilated loops, abolition of peristalsis, very thin wall (1 or 2 mm) in the arterial causes, and thickened and hypoechoic wall in the venous causes. In late cases, parietal microbubbles and flattening of the jejunal valvulae conniventes, fluid contents without gas, peritoneal effusion, portal gas [13, 14], or even hepatic abscesses and portal or mesenteric venous thrombosis have been described.

The superior mesenteric vein is often accessible (Fig. 6.14). Since it passes anterior to the rachis, it is possible to make a compression at this level in order to assess its patency, and without the help of the Doppler technique (see Chap. 12).

Small and Large Bowel: Other Acute Disorders

Pseudomembranous Colitis

Studying the ultrasound features of this complication of antibiotics may theoretically select the requirements for colonoscopy. The ultrasound pattern, insufficiently described in the literature [15], shows marked thickening of the colic wall, collapse of the lumen and frequent hemorrhagic ascites. Our rare observations also showed irregular debris floating within abundant intraluminal fluid, a pattern evoking parietal dissection.

Bowel Dilatation

The diagnosis is classically made using plain radiographs, which raises problems in the supine patient. CT is increasingly replacing plain radiographs. Yet ultrasound can be highly helpful when showing the following at the bedside:

- Dilatation of the bowel [16]. A dilated jejunum has a characteristic pattern (Fig. 6.15), but more subtlety is required to distinguish between dilated ileum and normal colon.
- Fluid content.
- Complete absence of wall and fluid content motion in the paralytic ileus, or sometimes to-and-fro movements only caused by the inertia of the sequestrated liquid.
- Peritoneal effusion is possible.
- An air–hydric level can be detected using the swirl sign. When the patient is supine and when the probe is applied vertically on the abdomen, a gas pattern is first observed. A slight pressure is then applied on the abdomen with the probe and free hand. When this pressure has shifted the gas collection, a fluid pattern immediately appears on the screen. At this moment, small movements made at the side of the bed will create swirls. The swirls result in sudden appearances and disappearances or an air pattern, with a complete acoustic barrier. Between the appearances of air, a fleeting image of fluid is visible (Fig. 6.16). This very suggestive pattern is of obvious meaning.

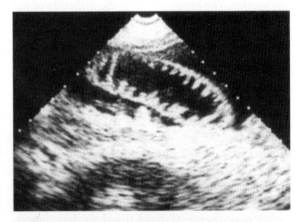

Fig. 6.15. Dilated jejunal loop. The wall, perfectly outlined between peritoneal effusion and fluid content, is thin. The fluid is here hypoechoic with hyperechoic particles. The caliper of this loop is 30 mm. Jejunal villi can be recognized (the fishbone sign). Small intestine occlusion. Transverse scan of the pelvic area

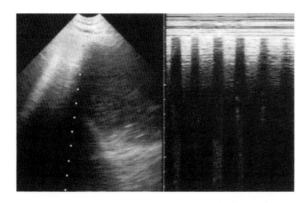

Fig. 6.16. Demonstration of the swirl sign using the time-motion mode. *Left*, real-time: air barrier at the left, fluid mass at the *right* of the screen. *Right*, time-motion: the air-fluid level has been gently shaken and the swirl created is the source of sudden transmissions of the ultrasonic beam

Fluid Digestive Sequestration

In a patient with shock, ultrasound detection of fluid sequestration within the intestines (Figs. 6.3, 6.5 and 6.9) immediately assumes a hypovolemic mechanism caused by digestive disorders (this sign will be associated with other ultrasound signs of hypovolemia). Briefly scanning the abdomen makes it possible to roughly evaluate the sequestrated volume of fluid.

In the same manner, in a patient with hemorrhagic shock, ultrasound can identify not yet exteriorized melena, which will appear as a fluid in the bowel (Fig. 6.17). This pattern is, of course, not specific. Nonetheless, ultrasound can thus logically be considered the first test able to detect GI tract hemorrhage, before the appearance of any clinical or biological anomaly.

Miscellaneous

Let us note here that the presence of peristalsis is as a rule a reassuring finding. In a series of 20 patients considered for emergency surgery, seven of them actually surgical cases, the sensitivity of an abolished peristalsis for the diagnosis of an abdominal disorder requiring prompt surgery was 100%, specificity 77% [10]. Consequently, in a suspicion of acute abdomen, the detection of a present peristalsis is a strong argument for ruling out a GI tract disorder requiring surgery.

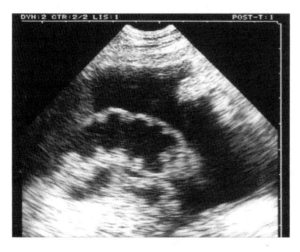

Fig. 6.17. Melena. This portion of the small bowel, outlined by ascites, is hypoechoic, indicating fluid. As was the case in this patient, this pattern can be the first sign of a GI tract hemorrhage

References

1. Schmutz GR, Valette JP (1994) Echographie et endosonographie du tube digestif et de la cavité abdominale. Vigot, Paris, p 16
2. Lim JH, Lee DH, Ko YT (1992) Sonographic detection of duodenal ulcer. J Ultrasound Med 11: 91–94
3. Weill F (1985) L'ultrasonographie en pathologie digestive. Vigot, Paris, pp 455–456
4. Lim JH, Ko YT, Lee DH, Lim JW, Kim TH (1994) Sonography of inflammatory bowel disease: findings and value in differential diagnosis. Am J Rœntgenol 163:343–347
5. Tuncel E (1990) Ultrasonic features of duodenal ulcer. Gastrointest Radiol 15:207–210
6. Smithius RHM and Op den Orth JO (1989) Gastric fluid detected by sonography in fasting patients: relation to duodenal ulcer disease and gastric-outlet obstruction. Am J Rœntgenol 153:731–733
7. Deutsch JP, Aivaleklis A, Taboury J, Martin B, Tubiana JM (1991) Echotomographie et perforations d'ulcères gastro-duodénaux. Rev Im Med 3:587–590
8. Teefey SA, Roarke MC, Brink JA, Middleton WD, Balfe DM, Thyssen EP, Hildebolt CF (1996) Bowel wall thickening: differentiation of inflammation from ischemia with color Doppler and duplex ultrasonography. Radiology 198:547–551
9. Danse EM, Van Beers BE, Goffette P, Dardenne AN, Laterre PF, Pringot J (1996) Acute intestinal ischemia due to occlusion of the superior mesenteric artery: detection with Doppler sonography. J Ultrasound Med 15:323–326
10. Lichtenstein D, Mirolo C, Mezière G (2001). L'abolition du péristaltisme digestif, un signe échographique d'infarctus mésentérique. Réanimation 10 [Suppl] 1:203

11. Fleischer AC, Muhletaler CA, James AE (1981) Sonographic assessment of the bowel wall. Am J Rœntgenol 136:887–891
12. Taboury J (1989) Echographie abdominale. Masson, Paris, pp 253–255
13. Kennedy J, Cathy L, Holt RN, Richard R (1987) The significance of portal vein gas in necrotizing enterocolitis. Am Surg 53:231–234
14. Porcel A, Taboury J, Aboulker CH, Bernod JL, Tubiana JM (1985) Aéroportie et infarctus mésentérique: intérêt de l'échographie. Ann Radiol 28:615–617
15. Downey DE and Wilson SR (1991) Pseudomembranous colitis: sonographic features. Radiology 180: 61–64
16. Mittelstaedt C (1987) Abdominal Ultrasound. Churchill Livingstone, New York

CHAPTER 7

Liver

The liver is the most voluminous plain organ, but is rarely a target for emergency therapeutic decisions in the ICU.

Mechanical ventilation, which lowers the diaphragm, can make its exploration easier. When the liver is located high, intercostal scans will be taken, provided the probe is small enough. Liver analysis is often not exhaustive in such conditions, but we will see that this limitation is relative in the critically ill patient.

Hepatomegaly

Although some operators can evaluate the weight of each lobe, the subjective feeling that the liver is enlarged is sufficient for others [1]. In the critically ill patient, it is more important to recognize the cause of this enlargement than the exact dimensions or weight. Usual causes in the ICU are acute right heart failure and cirrhosis.

The cardiac liver has a homogeneous structure, with dilatation of hepatic veins and vena cava inferior (Fig. 7.1). This finding will be accessory: the dilatation of the right heart and the lung disorder will then be recognized at the same time.

A cirrhotic liver will give numerous signs we will not detail here: a coarse pattern, a nodular pattern, atrophy or hypertrophy of one lobe with resulting global dysmorphia, absence of suppleness of the parenchyma, signs of portal hypertension (dilatation of the portal vein, ascites, reopening of the umbilical vein, splenomegaly and others). See Fig. 6.7, p 35, for an illustration of esophageal varices.

As regards tumoral or infectious (abscesses) enlargements, the cause will immediately appear on the screen.

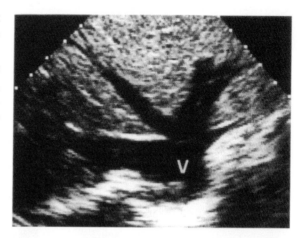

Fig. 7.1. Liver in right heart failure. Dilatation of the three hepatic veins, which open into an inferior vena cava (*V*) also dilated. Note that this scan does not reflect the site where its caliper should be measured (see Chap. 13, p 82). Epigastric subtransverse scan

Portal Gas

This is a situation where ease of diagnosis and efficiency of therapeutic management meet. Portal gas generally requires prompt surgery [2, 3]. In a critical scene, portal gas immediately evokes mesenteric infarction. Ultrasound may give a chance for the patient to benefit from an earlier diagnosis. Portal gas is traditionally considered a pejorative sign [4], but this feeling is based on radiographic findings. Yet ultrasound is more sensitive than radiographs [2]. In addition, we have seen surgical success even when ultrasonic portal gas was present.

Portal gas yields numerous punctiform hyperechoic images without acoustic shadow within the liver parenchyma and usually peripheral (Fig. 7.2). In this case, we speak of static portal gas. In some cases, one can observe a flux of gas particles at the portal vein (Fig. 7.3), a sign we called dynamic portal gas. In these cases, when such particles are seen coming from the superior mesenteric vein and not

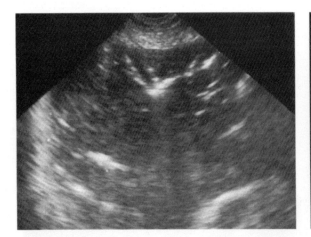

Fig. 7.2. Static portal gas. Numerous hyperechoic punctiform opacities, without acoustic shadow, within the liver of a patient with mesenteric infarction. Note that this patient survived, in spite of the classically poor prognosis of portal gas

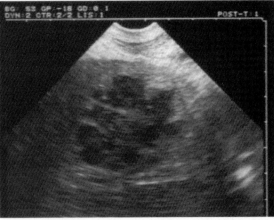

Fig. 7.4. Hepatic abscess (*Klebsiella*). Hypoechoic heterogeneous mass within the hepatic parenchyma

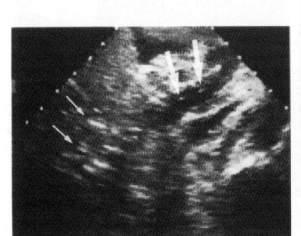

Fig. 7.3. Dynamic portal gas. A visible flow with hyperechoic particles (*large arrows*) is observable in the portal vein. Static portal gas can be seen (*small arrows*). Oblique scan of the right hypochondrium, in the axis of the portal vein (*large arrows*), in a patient with septic shock

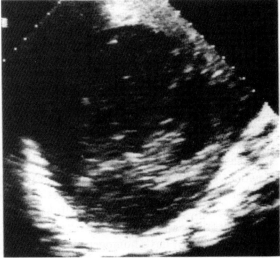

Fig. 7.5. Hepatic abscess (*Streptococcus milleri*). Huge round hypoechoic mass. In real-time, this mass had a characteristic internal motion, which indicated a fluid nature. Percutaneous ultrasound-guided drainage (see Fig. 26.1, p 173) has withdrawn 1,150 cc of frank pus

from the splenic vein, they originate logically from the GI tract.

Volvulus or strangulation, ulcerous colitis, and intra-abdominal abscesses are other causes described in the adult [4].

Hepatic Abscess

Ultrasound is a quick and user-friendly method of diagnosis, since it spares the highly unpleasant pain caused by liver shaking. Pain is often absent in a encephalopathic patient in shock, hence the interest of a systematic ultrasound examination in any critically ill new arrival.

Abscess yields an image contrasting with the regular hepatic echostructure. It is generally hypoechoic, heterogeneous, and roughly round (Fig. 7.4). A very characteristic sign is sometimes observed: within the mass, an internal movement is visible, in rhythm with respiration. This is in fact the inertia of the pus caused by the movement (Fig. 7.5), the equivalent of the plankton sign discussed in Chap. 5. In our observations, it proves the fluid nature of the

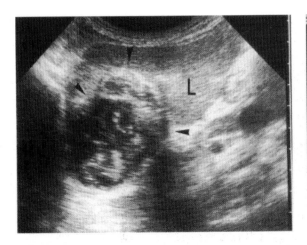

Fig. 7.6. Hydatid cyst of the liver (*arrowheads*). The heterogeneous pattern indicates complication, here suppuration, which was confirmed at the laparotomy of this patient in septic shock. Longitudinal scan of the liver. *L*, liver

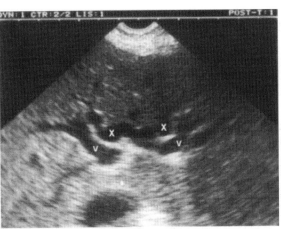

Fig. 7.8. Dilatation of intrahepatic bile ducts. Vessels (*X*) are visible anterior to portal bifurcation (*V*), producing a double channel pattern

Diffuse Infectious Disorders

Tuberculous hepatic miliary can be missed by ultrasound (Fig. 7.7). In cases where there is strong clinical suspicion, a prompt liver biopsy should provide bacteriological confirmation.

Cholestasis

Ultrasound is a quick and simple way to check for the normal condition of the bile ducts. However, cholestasis occurring in a ventilated patient is very frequent. In our observations, the cause of cholestasis is always medical: sepsis or impairment of venous return. We are still awaiting a surgical cause of cholestasis in a patient initially ventilated for another reason.

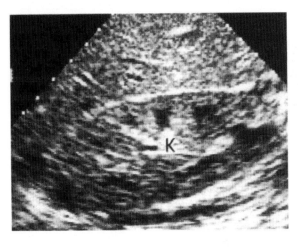

Fig. 7.7. Diffuse tuberculous miliary. In this longitudinal scan of the liver and the kidney (*K*), it is hard to detect frank anomalies. Real-time showed that the liver parenchyma pattern was homogeneously granular, but one can consider it is a subtle sign

collection (regardless of the presence or absence of posterior enhancement), and above all it indicates pathological fluid (pus, blood). Highly echoic images are sometimes seen, indicating microbial gas. Pleural effusion (generally radiopaque) is possible.

Amebic abscess yields a hypoechoic, well-limited collection.

Hydatidosis should be evoked before any puncture of fluid hepatic mass. This does not cause a problem when the cyst is well defined and anechoic, since there is no emergency, but it may in the suppurative forms, when the cyst becomes echoic and heterogeneous (Fig. 7.6).

This said, in case of an obstacle, ultrasound will detect bile duct dilatation: the intrahepatic duct anterior to the portal bifurcation (Fig. 7.8) or the main duct anterior to the portal vein (Fig. 7.9). The normal caliper of the main bile duct is said to be 7 mm (up to 12 mm in the case of an old cholecystectomy), but some authors have fixed the upper limit at 4 mm [5]. When the common bile duct is dilated, it acquires a tortuous route and cannot be visualized in a single view. The sensitivity of ultrasound is poor for detection of common bile duct calculi, which rarely produce posterior shadows, even if massive [6].

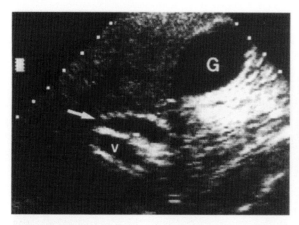

Fig. 7.9. Anterior to the portal vein (*V*), the common bile duct (*arrow*) is dilated with a 9-mm caliper. Oblique scan of epigastric area. *G*, gallbladder

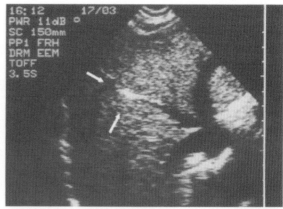

Fig. 7.10. Hyperechoic structure, highly dynamic in real-time, visible at the median hepatic vein (*arrows*). Trapped air in the hepatic venous system. Subtransverse epigastric scan acquired with an Ausonics 2000 device

Hepatic Vein Disorders

Ultrasound is an excellent noninvasive method for examining hepatic vein disorders [7]. In the Budd-Chiari syndrome with hepatic veins thrombosis, these veins are filled with echoic material, are filiform, or are not visible if they have the same echogenicity as the liver. Other signs exist but their description would deviate too far from our initial objectives. Faithful to a maximal use of two-dimensional ultrasound, and regarding the rarity of this disorder (at least in our institutions), we think that two-dimensional ultrasound should be done first. Visualization of anechoic hepatic veins, which can be compressed with the pressure of the probe, indicate patency of these veins. Obviously, the operator should search for more frequent diseases to explain the symptoms bringing suspicion of Budd-Chiari syndrome. If the examination remains noncontributory, then and only then should a Doppler study be indicated.

In critically ill patients, mobile gas is sometimes observed in the median and left hepatic veins, which are the non-declive veins (Fig. 7.10). The most logical explanation is that air accidentally coming from perfusions (in the arms, for instance) are trapped in these veins. A tricuspid regurgitation, very frequent in the mechanically ventilated patient, may be the cause.

Hepatic Tumors

Recognition of metastases may give a theoretical element of prognosis in the acute phase. They are usually known, but they can be discovered by ultrasound when no anamnesis is available. The pattern is usually characteristic: multiple disseminated images with anarchic distribution, isoechoic, or hyperechoic with a fine hypoechoic stripe, or again hypoechoic images (Fig. 7.11). As regards other tumors, we will be brief, since they do not need particular treatment or reflexion during the stay in the ICU. A round, regular, anechoic image is generally a biliary cyst, sometimes also an uncomplicated hydatid cyst. An echoic heteroge-

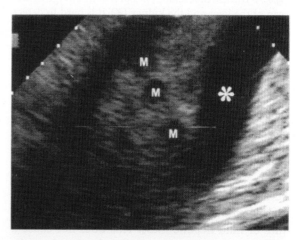

Fig. 7.11. Hypoechoic masses, disseminated in the liver with a multicentric pattern. Hepatic metastases. Peritoneal effusion surrounding the liver (*asterisk*) secondary to peritoneal metastases

neous mass within a cirrhotic parenchyma will be suggestive of hepatocarcinoma. These tumors, and others (adenoma, focal nodular hyperplasia, angioma, primitive malignant tumors, heterogeneous steatosis, etc.) are extensively described in excellent textbooks [1, 8, 9].

Miscellaneous

In hepatic trauma, identifying hemoperitoneum is possible, as well as direct patterns of liver contusion in favorable cases (see Fig. 24.1, p 165).

Aerobilia can be pathological, in ileus by impacted gallstone, or physiological, after biliary surgery. Numerous air opacities are visible along the biliary vessels, which converge to the hilum. Thus, the images are more central than in portal gas.

Interventional Ultrasound

Percutaneous Aspiration or Drainage of Liver Abscess

We were able to successfully aspirate hepatic abscesses with the material described in Chap. 26. Deep locations or locations near the dome can cause technical problems.

Percutaneous or Transjugular Liver Biopsy

The presence of permanent ultrasound assistance means that emergency liver biopsies can be carried out. Three indications can be imagined in the ICU:

- Documenting diffuse tuberculosis before treatment
- Proving the malignant nature of liver images, if this finding can modify immediate treatment
- Investigating fulminant hepatitis.

In this last case, hemostasis disorders usually require a transjugular approach, which usually means transportation of a critically ill patient to a specialized center. Yet transjugular hepatic biopsy could be performed at the bedside under sonographic guidance, with a double impact. First, immediate and successful catheterization of the internal jugular vein (see Chap. 12); second, after insertion of the material, guidance toward the target. Let us specify that radioscopy, which gives a good overview, creates irradiation, and above all, reduces a three-dimensional shape (the liver) to a two-dimensional image. Two-dimensional ultrasound, in well-trained hands, gives a three-dimensional image of an area. It accurately steers the material through the inferior vena cava, then the hepatic veins. This visual guidance should decrease the number of incidents that occur with radioscopic guidance.

References

1. Menu Y (1986) Hépatomégalies. In: Nahum H, Menu Y (eds) Imagerie du foie et des voies biliaires. Flammarion, Paris, p 86–96
2. Lee CS, Kuo YC, Peng SM et al (1993) Sonographic detection of hepatic portal venous gas associated with suppurative cholangitis. J Clin Ultrasound 21: 331–334
3. Traverso LW (1981) Is hepatic portal venous gas an indication for exploratory laparotomy? Arch Surg 116:936–938
4. Liebman PR, Patten MT, Manny J (1978) Hepatic portal veinous gas in adults. Ann Surg 187:281–287
5. Berk RN, Cooperberg PL, Gold RP, Rohrmann CA Jr, Ferrucci JT Jr (1982) Radiography of the bile ducts. A symposium on the use of new modalities for diagnosis and treatment. Radiology 145:1–9
6. Weill F (1985) L'ultrasonographie en pathologie digestive. Vigot, Paris
7. Menu Y, Alison D, Lorphelin JM, Valla D, Belghiti J, Nahum H (1985) Budd-Chiari syndrome, ultrasonic evaluation. Radiology 157:761–764
8. Taboury J (1989) Echographie abdominale. Masson, Paris
9. Weill F (1985) L'ultrasonographie en pathologie digestive. Vigot, Paris, pp 455–456

Chapter 8

Gallbladder

Acute acalculous cholecystitis, a classic complication of the critically ill and a classic indication for general ultrasound, deserves an entire chapter. Our experience suggests two comments. First, this disorder seems to affect mostly the surgical patient and is exceptional in the medical ICU. Second, if ultrasound can accurately describe many data, the very interpretation of these data remains subtle. In fact, the gallbladder can have a vast variety of patterns, from the normal to the pathological, in passing by the picturesque (Figs. 8.1, 8.2). A strictly normal gallbladder in the ICU is an infrequent finding (see Fig. 4.7, p. 21). The variations in volume, wall thickness, content, shape and surroundings can create infinite combinations. Some are variants of the normal, some are pathological but do not require emergency procedures, and others need prompt surgery, beneficial in slightly less than half of the cases in our experience.

Classic Signs of Acute Acalculous Cholecystitis

Acute acalculous cholecystitis is found in 5%–15% of acute cholecystitis and 47% of postoperative cholecystitis [1]. The diagnosis is based on infectious syndrome and local signs in an exposed patient [2]. Histology alone provides definite diagnosis, a mandatory sign being wall infiltration by neutrophils. Ultrasound patterns classically associate:

- Enlarged gallbladder, with a long axis caliper over 90 mm and a short axis over 50 mm.
- Wall thickening greater than 3 mm.
- Sludge (echoic, compact, declive sediment).
- Perivesicular fluid collection, valuable in the absence of ascites.
- Murphy's sign: pain due to the pressure of the gallbladder. Since ultrasound has the merit of precisely locating the gallbladder, ultrasound identification of Murphy's sign is mentioned when the probe itself applied in front of the gallbladder creates elective pain.

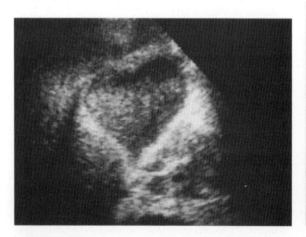

Fig. 8.1. Elegance is not forbidden in an organ as critical as the gallbladder. A simple folding at the hepatic aspect is enough to confer this discrete charm

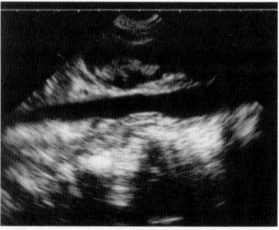

Fig. 8.2. In another gallbladder, a very irregular sludge seems to represent a crouched coyote in an asymptomatic patient

Sensitivity of ultrasound is weak (67%) for some [3], high (90%–95%) for others [4, 5]. When distension, thickening and sludge are combined, sensitivity falls, but specificity climbs [2].

Observations of Acute Acalculous Cholecystitis

Acute acalculous cholecystitis seems to be specific to the surgical ICU. It seems to happen especially after major vascular surgery such as aorta surgery. Although ultrasound can localize the gallbladder and can accurately delineate the phenomena described above, we suspect that these signs, taken one after another or even together, are subject to a problem of interpretation. Our observations of histologically proven acute acalculous cholecystitis have led to the following conclusions (Fig. 8.3).

Size

On average, the gallbladder measured 103 mm on the long axis (range, 65–150 mm) and 40 mm on the short axis (range, 29–55 mm).

The Wall

The wall was always moderately thickened, measuring on average 4.6 mm (minimum observed, 3.0 mm; maximum, 6.2 mm).

Sludge

Sludge was present in 90% of cases.

Murphy's Sign in Ultrasound

We observed a genuine Murphy sign in 8% of cases.

Perivesicular Effusion

We observed selective effusion in 12% of cases.

The problem begins with the existence of a disorder very frequently encountered in our histology reports: chronic subacute cholecystitis. This disorder will raise serious diagnostic problems.

Chronic Subacute Cholecystitis

Chronic subacute cholecystitis is a histological definition. In fact, neither ultrasound nor even perioperative findings can distinguish it from the acute acalculous cholecystitis (Fig. 8.4). Nearly half of our

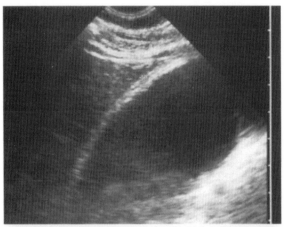
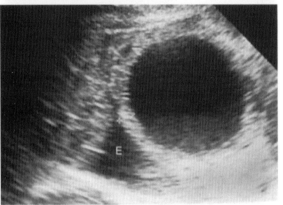

Fig. 8.3a, b. Acute acalculous cholecystitis, with histological proof. A homogeneous thickening of the wall (4 mm), a caliper of 30 mm, and dependent sludge are depicted. There was no pain in this sedated patient. Above all, this gallbladder is suspect because the patient developed fever after major aortic surgery. **a** Longitudinal scan. **b** Transverse scan, in which a moderate peritoneal effusion is visible (*E*)

patients operated for suspicion of acute acalculous cholecystitis in fact had chronic subacute cholecystitis. Chronic subacute cholecystitis does not seem to require surgery. In our observations, the average long axis was 105 mm (range, 84–160 mm), average caliper, 37 mm (range, 23–56 mm), average wall thickness, 4.5 mm (range, 3.0–7.0 mm), sludge was present in 66% of cases, Murphy's sign in 10% and localized effusion was never present.

However, these data are quite similar to those seen in acute acalculous cholecystitis (Table 8.1). One consequence is that this disorder is diagnosed, with subsequent surgery, with the same frequency as acute acalculous cholecystitis. This probably means useless surgery, in other words, increased operative risk, and above all, this means that the initial problem remains undiagnosed. A

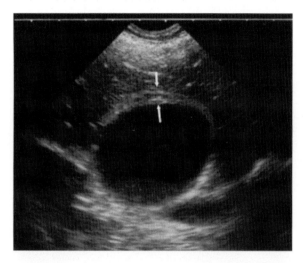

Fig. 8.4. This gallbladder has a homogeneous 5.5-mm thickened wall, a pattern not really different from Fig. 8.3. Sludge is also discretely present. Pathological examination confirmed the diagnosis of chronic subacute cholecystitis

perioperative pattern is sometimes misleading, and many gallbladders considered acute or even gangrenous become simple chronic subacute cholecystitis once under the microscope.

In addition, we have frequently seen ultrasonically suspect gallbladders that were not operated and that spontaneously normalized. The problem is again intricate, as some authors argue that certain acute acalculous cholecystitis cases can be cured without surgery, but this should be proven with a solid methodology.

Common Gallbladder Patterns Seen in the Intensive Care Unit

It may be timely to specify one point here. In our experience, mostly from the medical ICU and over a systematic observation of our patients since 1989, acute acalculous cholecystitis was rare. In 11 years of practice in the medical ICU and 5 years in the surgical ICU (with major vascular surgery), we found one case of acute acalculous cholecystitis every 500 days of physician presence in medical patients and 23 days for surgical patients. This means a frequency 20 times lower for medical patients.

In our critically ill patients with a stable status and with no superimposed clinical problem, the majority of gallbladders were enlarged and contained sludge. Wall thickening was extremely frequent; the major form of this thickening will be dealt with in a later section. Peritoneal effusion was routine in severely critically ill patients. Let us examine these signs in detail.

Volume

Volume can vary between complete vacuity to distension. A completely empty gallbladder can be hard to detect. One should follow precise landmarks: the right branch of the portal bifurcation leads to the fossa vesicae felleae, which always leads to the gallbladder space (Fig. 8.5). An empty gallbladder is, in principle, functional, since it is able to contract. It may also be perforated. A distended gallbladder (long axis >90 mm, short axis >50 or 40 mm) is the rule in patients under parenteral feeding and taking morphines (Fig. 8.6). The lumen can be virtual and the wall thickened (Fig. 8.7). Among other patterns, one can see septate contents, variations in length, complete calcifications of the wall, or tumors. Images of these anomalies are accessible in abdominal ultrasound textbook [6, 7].

Wall Thickening

The normal wall measures between 1.5 and 3 mm. A 4-mm cut-off has the advantage of being reliable, but this notion may be obsolete. Modern units have an improved definition, and wall thick-

Table 8.1. Acute acalculous versus chronic subacute cholecystitis

	Acute acalculous cholecystitis	Chronic subacute cholecystitis
Wall thickening	4.6 mm (3.0–6.2)	4.5 mm (3.0–7)
Long axis	103 mm (65–150)	105 mm (84–160)
Short axis	40 mm (29–55)	37 mm (23–56)
Sludge	90%	66%
Localized perivesicular effusion	12%	0
Murphy's ultrasound sign	8%	10%

Extreme values are in parentheses.

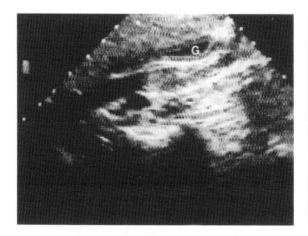

Fig. 8.5. Example of an empty gallbladder. This discrete image should be recognized to avoid erroneous diagnoses

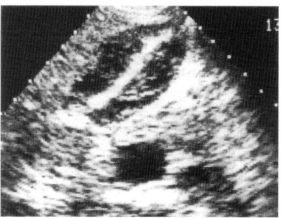

Fig. 8.7. This gallbladder has virtual lumen, reduced to an echoic stripe, and an extremely thickened wall, to 12 mm. Laparotomy and pathology revealed simple gallbladder edema in this patient in septic shock with major lung injury

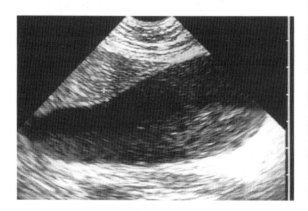

Fig. 8.6. This enlarged gallbladder (100×40 mm) has a thickened wall (3.6 mm) and roughly 40% sludge, which is very frequent in the ICU. However, this gallbladder did not provoke symptoms in a female patient admitted for ARDS (aspiration pneumonia), who eventually recovered

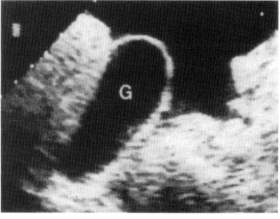

Fig. 8.8. The wall of this gallbladder is perfectly outlined between bile (*G*) and ascites. This wall is perfectly fine, a pattern which easily invalidates the traditional idea that ascites causes gallbladder wall thickening

ening greater than 3 mm should be considered with care. The wall can be very distinct when outlined between bile and peritoneal effusion (Fig. 8.8). It can be impossible to measure precisely. In some cases, there is no contrast between the gallbladder wall and hepatic parenchyma, which makes any exact measurement illusory.

We routinely find a thickened wall (Fig. 8.6). It can be split, with two echoic layers surrounding an hypoechoic layer. A striated pattern is described as a sign of acute acalculous cholecystitis [8], but the follow-up of our patients does not support this impression.

Traditionally, a thickened wall is nearly equivalent to acute acalculous cholecystitis. Experience shows that this sign has very low specificity. The classic list of causes includes ascites, hepatitis, hypoalbuminemia, and cardiac failure, a rather vague term [9]. Observation shows that, in the case of ascites, and in spite of the traditional widespread belief to the contrary the wall can be perfectly thin (see Fig. 8.8). Cardiac failure is an overly vague notion. In contrast, acute right heart failure should certainly be considered a prominent cause, so much so that we speak of cardiac gallbladder (see next section).

Sludge

Sludge is nearly always present in the critically ill patient, since the gallbladder does not work in a physiological way. The pattern can vary greatly, although we could not attribute a particular value to each. Sludge can be homogeneous (Fig. 8.6) or heterogeneous, containing hyperechoic dots (could microlithiases be included in the mass?). The interface between the sludge and the anechoic nondependent bile can be regular (Fig. 8.6) or ragged (Fig. 8.2). Sludge can be discrete or massive: in some cases, a 100% sludge yields a pattern isoechoic to the liver – a hepatization of the gallbladder, so to speak (Fig. 8.9). Excellent knowledge of anatomy is then required to recognize the gallbladder. The sludge can be tumor-shaped. Last, sludge appears at variable stages: usually occurring during a prolonged stay, it can be present at admission. Eventually, it can completely vanish.

Murphy's Sign in Ultrasound

Murphy's sign is very rarely contributive since critically ill patients are all sedated or, if not, they are in shock or encephalopathic. Pain is either absent or diffuse to the entire body.

Peripheral Peritoneal Effusion

Peritoneal effusion is very frequent in the critically ill patient. Localized effusion in acute cholecystitis is a rare finding. Moreover, the very routine observation of thin-wall gallbladders surrounded by extensive peritoneal effusion will prove to any operator that peritoneal effusion is not in itself a cause explaining wall thickening.

In Summary

In conclusion, all these changes are routine and of little relevance, even when integrated in a suggestive context.

A Distinctive Feature: Major Wall Thickening of the Cardiac Gallbladder

We regularly and frequently observe gallbladders with the remarkable feature of major wall thickening, more than 7 mm, up to 18 mm (Fig. 8.10). This pattern:

- Always occurs in patients with right heart failure such as acute asthma, pneumonia, adult respiratory distress syndrome (ARDS), pulmonary embolism, acute tricuspid regurgitation, exacerbation of chronic obstructive pulmonary disease (COPD), in the most severe forms. This population is more often seen in medical ICUs, hence possibly a higher rate of cases observed here.
- There is no local sign in these generally sedated patients.

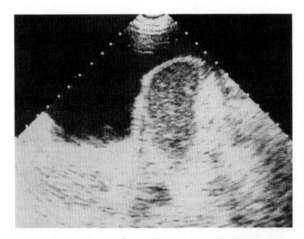

Fig. 8.9. This gallbladder, which seems to be floating within massive peritoneal effusion, contains a totally echoic lumen. This shows complete sludge in an asymptomatic patient

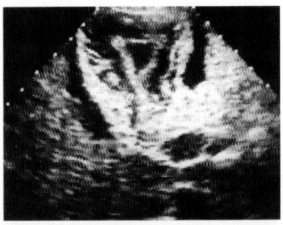

Fig. 8.10. Cardiac gallbladder. The wall of this gallbladder is extremely enlarged, up to 20 mm. A hypoechoic layer is surrounded by two echoic layers. The lumen is narrow, probably because of the space taken by the walls. This patient has acute right heart failure. Pathology confirmed simple wall edema

- The gallbladder cavity itself is often small, possibly because the walls enlarge to the detriment of the cavity.
- In our experience, a dozen observations among a large number were positively documented, using laparotomy, for instance. All of these observations were the result of wall edema, sometimes chronic subacute cholecystitis, but never acute acalculous cholecystitis.
- Time allowing, one can observe the complete regression of this major thickening.

We suggest labeling this frequent observation of overly thickened wall the cardiac gallbladder, with analogy to cardiac liver or cor pulmonale. It can be assumed that the cardiac gallbladder:

- Is above all the manifestation of congestive phenomena that are observable at the gallbladder wall, since this is an accessible area, as retinal vessels are a privileged site to assess general circulatory function.
- Is frequent.
- Can be occult, because this is a transitory feature.

Conversely, an ultrasound examination performed at the climax of the wall thickening can lead to an erroneous diagnosis of acute acalculous cholecystitis, and result in a certain number of unnecessary laparotomies.

There is a clinical relevance to the recognition of cardiac gallbladder. Data suggest that the detection of thickening over 7 mm in a patient with symptoms that may evoke acute acalculous cholecystitis should incite the physician to search for another cause to explain the present symptoms. A laparotomy would not only be useless, but also deleterious if the real cause is not recognized.

How to Establish the Diagnosis of Acute Acalculous Cholecystitis

In conclusion, we believe that if ultrasound is an excellent method for localizing and measuring the gallbladder, it cannot distinguish the surgical emergency from an insignificant variant of the normal.

Patient Background and Current Situation

It seems wise to evoke acute acalculous cholecystitis only in well-defined patients. A major vascular surgery (of the aorta, for example) is found in half

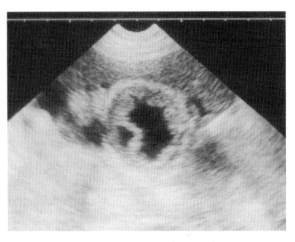

Fig. 8.11. The gallbladder of this patient admitted for exacerbation of chronic respiratory disease had a very unsettling pattern: a scalloped wall with possible debris detached from the left aspect. Pathology authenticated a simple chronic subacute cholecystitis

of our cases, a major trauma in a quarter of cases. As for chronic subacute cholecystitis, major vascular surgery is found in only 16% of cases, trauma in 33%. Almost all patients with cardiac gallbladder have ARDS or multiple organ failure.

Considering Certain Ultrasound Signs

Let us recall the conclusions of the previous section: a wall thickening greater than 7 mm in a medical ICU patient suspected of having acute acalculous cholecystitis should prompt a search for another cause explaining the symptoms. We still find this policy valuable after 12 years of observations.

A subtle study of the signs at the wall showing ulcerations would be valuable, but our investigations are at a standstill. We sometimes thought we had visualized shreds detached from the mucosa (Fig. 8.11), but laparotomy and pathology ruled out the diagnosis of acute acalculous cholecystitis. Detachment of the mucosa with shreds floating in the lumen is described in the literature as a sign of gangrenous cholecystitis [10]. In acute acalculous cholecystitis, there is the notion of a very thin wall in a preperforative stage. It seems therefore wise to study the wall in its entirety, screening for areas of weakness. However, we are still awaiting our first case.

Intramural gas should be observed in emphysematous cholecystitis. We have not had the privilege of observing this sign, probably rare. Mural gas should give hyperechoic punctiform images, a sign which should not be confused with cholesterol calculi contained in the Rokitansky-Aschoff sinuses within the delightful setting of gallbladder adenomyomatosis, although this is of little interest to us here.

Other Tests

Doppler

If the Doppler could accurately distinguish between ischemic and edematous wall, it would then be potentially of interest.

CT

CT does not contribute a great deal, since a careful ultrasound is almost always able to analyze the gallbladder. Let us note here that measurement of wall thickening is much more accurate using ultrasound rather than CT [11]. As a rule, and not only at the gallbladder level, ultrasound has a focal resolution superior to that of CT (see Fig. 8.12).

Dynamic Cerulein Test and Scintigraphy

These two tests are of little value [10].

Ultrasound-Guided Aspiration of Gallbladder Bile

In our experience, this procedure is extremely simple, as long as basic rules are respected. A simple 21-gauge needle is sufficient. The gallbladder must be punctured throughout the liver (the hole will be recovered by the liver). Bile leakage cases described in the literature result from transperitoneal approaches. The dependent bile is aspirated, since the nondependent area may yield false-negatives. Since pathological bile is viscous, aspiration must be vigorous. The amount of aspired bile should be sufficient to diminish the possible hyperpressure and thus limit the risk of leakage. Conversely, if percutaneous drainage is envisaged, the volume of the gallbladder should not be decreased too much. When the tap is in place, the needle is withdrawn and strong manual compression is applied at the point of puncture. If strong compression is not applied, for fear of bile leakage, hemoperitoneum or subcapsular hematoma of the liver can result in patients with impaired hemostasis. Control at 1 and 12 h will search for perivesicular effusion. The vesicular bile of a critically ill

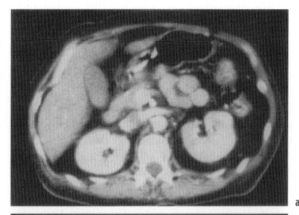
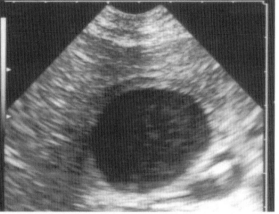

Fig. 8.12a, b. It is not difficult to objectify ultrasound's superiority (**b**) over CT (**a**) as regards focal spatial resolution. The gallbladder wall, difficult to view on CT, is sharply visible on ultrasound and can accurately be measured

patient is usually dark brown or green brown, mildly sticky, sometimes black like tar, and viscous, when the sludge itself has been aspirated.

The risk of vesicular tap is possible though rare. It should be compared with the risk of allowing angiocholitis or cholecystitis to develop, which can be clinically difficult to detect. Of 25 procedures performed as described, we have encountered no complications.

This technique is simple and seems safe. But is it relevant? For some, it is contributive [12] when it provides proof of infection at the bedside, which should be present in 66% of the cases [13]. Other studies [14] question the sensitivity of this procedure, almost always performed in patients taking antibiotic therapy. Leukocytes found in the gallbladder bile should indicate cholecystitis [14]. The most important limit is that acute acalculous cholecystitis appears more as an ischemic process than an infectious one [15].

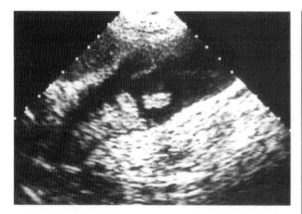

Fig. 8.13. Acute purulent cholecystitis. Dependent masses (lower part of the image) are not typical of sludge since they are rather echoic, nor do they evoke calculi, since there is no posterior shadow. Images of membranes seeming to detach from the wall are visible at the upper part of the image. An ultrasound-guided tap immediately confirmed the diagnosis (frank pus) and the patient was immediately sent to the operating room

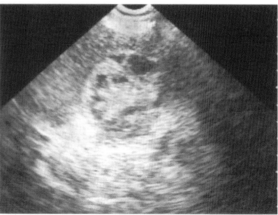

Fig. 8.14. Gallbladder space hematoma. Heterogeneous echoic pattern, often found in the gallbladder space after surgical removal

We have had one case where diagnosis of acute infectious cholecystitis was immediately made in a patient admitted for shock, thanks to bedside ultrasound-guided aspiration of gallbladder bile. It is true that the gallbladder had an atypical pattern, with particularly echoic sludge, but this pattern could have been considered as a variant of normal (one more to add to a long list). Yet the puncture had withdrawn frank pus, and the patient was rightly sent to the operating room (Fig. 8.13).

Other Pathological Patterns of the Gallbladder

Cholecystectomy Space

Infection of the cholecystectomy space is frequently suspected (Fig. 8.14). Ultrasound-guided aspiration appears to be an accessible procedure and can distinguish pus retention from old sterile blood.

Calculous Acute Cholecystitis

This disorder is rarely of interest to the intensivist. The calculi give a dependent hyperechoic, round image with frank posterior shadow (see Fig. 1.4, p 6). Calculi are frequently observed and should be respected if quiet. Obviously, the smaller the calculi, the more they are able to move and cause trouble. The association of calculi, thickened wall and Murphy's sign on ultrasound has a positive predictive value of 95%, and the absence of these three signs has a negative predictive value of 98% [16]. Acute calculous cholecystitis rarely raises diagnostic problems.

Acute Acalculous Cholecystitis in Calculous Gallbladder

Since calculi are a frequent finding in the general population, one should find a pertinent term to label an acute cholecystitis of critically ill patients occurring in a previously calculous gallbladder.

Interventional Ultrasound

Diagnostic Aspiration of Bile

This procedure has been discussed in »Ultrasound-Guided Aspiration of Gallbladder Bile.«

Percutaneous Cholecystostomy

Some authors underline the easiness of this procedure and the low rate of complications [14, 17]. Technical requirements are the same as those described for aspiration. Kits are available, with laterally perforated pigtail catheters. They normally prevent parietal perforation and dislocation of material. A series of 322 procedures described a null mortality rate and a morbidity rate of 2%–5% [18]. This procedure [19] was advocated as an alternative to surgery in the critically ill [17, 20]. It provides relief of an obstacle located in the biliary tract. It was even shown to be effective in sepsis without obvious causes [14]. Other teams mistrust this apparently attractive technique, since a fragile wall can easily be perforated [15]. We add two major arguments against this procedure: first, histological proof is unavailable, and no conclusion can be drawn from how the situation evolves. Second, acute acalculous cholecystitis seems more an ischemic than an infectious disorder, and this indicates that the gallbladder, and not its content, should be removed.

From a methodological point of view, it would be valuable to study a population with clinical and ultrasound patterns suggestive of acute acalculous cholecystitis, and to compare the progression of operated patients and those with a spontaneous recovery. Such a study will be hard to conduct since it is ethically difficult to take the risk of allowing a genuine acute cholecystitis to evolve [21]. Note simply that this methodological shortcoming weighs heavily in the published studies [14].

References

1. Cooperberg PL and Gibney RG (1987) Imaging of the gallbladder, state of the art. Radiology 163:605-613
2. Bodin L and Rouby JJ (1995) Diagnostic et traitement des cholécystites aiguës alithiasiques en réanimation chirurgicale. ACTUAR 27:57-64.
3. Shuman WP, Rogers JV, Rudd TG, Mack LA, Plumley T, Larson EB (1984) Low sensitivity of sonography and cholescintigraphy in acalculous cholecystitis. Am J Roentgenol 142:531-537
4. Mirvis SE, Vainright JR, Nelson AW, Johnston GS, Shorr R, Rodriguez A, Whitley NO (1986) The diagnosis of acute acalculous cholecystitis: a comparison of sonography, scintigraphy and CT. Am J Roentgenol 147:1171-1179
5. Van Gansbeke D, Matos C, Askenasi R, Braude P, Tack D, Lalmand B, Avni EF (1989) Echographie abdominale en urgence, apport et limites. Réanimation et Médecine d'Urgence. Expansion Scientifique Française, Paris, pp 36-53
6. Nahum H and Menu Y (1986) Imagerie du foie et des voies biliaires. Flammarion, Paris
7. Weill F (1985) L'ultrasonographie en pathologie digestive. Vigot, Paris
8. Teefey SA, Baron RL, Bigler SA (1991) Sonography of the gallbladder: significance of striated thickening of the gallbladder wall. Am J Roentgenol 156:945-947
9. Slaer WJ, Leopold GR, Scheible FW (1981) Sonography of the thickened gallbladder wall: a non-specific finding. Am J Roentgenol 136:337-339
10. Chagnon S, Laugareil P, Blery M (1988) Aspect échographique de la lithiase biliaire et de ses complications locales. Feuillets de Radiologie 28:415-423
11. Bodin L, Rouby JJ, Langlois P, Bousquet JC, You K, Viars P (1986) Cholécystites aiguës alithiasiques en réanimation. Etude randomisée comparant 2 méthodes thérapeutiques: chirurgie et ponction drainage percutanée sous contrôle échographique. In: Viars P (ed) Actualités en Anesthésie-Réanimation. Arnette, Paris, pp 157-167
12. McGahan JP, Walter JP (1985) Diagnostic percutaneous aspiration of the gallbladder. Radiology 155:619-622
13. Sicot C (1992) Les cholestases intra-hépatiques aiguës chez les malades de réanimation. Réan Urg 1:578-583
14. Lee MJ, Saini S, Brink JA, Hahn PF, Simeone JF, Morrison MC, Rattner D, Mueller RP (1991) Treatment of critically ill patients with sepsis of unknown cause: value of percutaneous cholecystostomy. Am J Roentgenol 156:1163-1166
15. Langlois P, Bodin L, Bousquet JC, Rouby JJ, Godet G, Davy-Mialou C, Wiart D, Cortez A, Chomette G, Grelet J, Chigot JP, Mercadier M (1986) Les cholécystites aiguës non lithiasiques post-agressives. Apport de l'échographie au diagnostic et au traitement dans 50 cas. Gastroenterol Clin Biol 10:238-243
16. Ralls PW, Colletti PM, Lapin SA, Chandrasoma P, Boswell WD, Ngo C, Radin DR, Halls JM (1985) Real-time sonography in suspected acute cholecystitis. Radiology 155:767-771
17. Vogelzang RL, Nemcek Jr AA (1988) Percutaneous cholecystostomy: diagnostic and therapeutic efficacy. Radiology 168:29-34
18. Malone DE (1990) Interventional radiologic alternatives to cholecystostomy. Radiol Clin North Am 28:1145-1156
19. Roche A, Cauquil P, Houlle D (1986) Radiologie interventionnelle des voies biliaires. In: Duvauferrier R, Ramee A, Guibert JL (eds) Radiologie et échographie interventionnelles, tome 2. Axone, Montpellier, pp 457-494
20. Picus D (1995) Percutaneous gallbladder intervention. Eur Radiol 5 [Suppl]:S180
21. Johnson LB (1987) The importance of early diagnosis of acute acalculous cholecystitis. Surg Gynecol Obstet 164:197-203

CHAPTER 9

Urinary Tract

The study of the urinary tract is vast, since mechanical, infectious and hemodynamic phenomena are all involved.

The normal pattern of the kidney and bladder is described in Chap. 4 (Figs. 4.8 and 4.9, pp 21–22).

Renal Parenchyma

The diagnosis of acute renal failure is biological, and the main advantage of ultrasound is first to rule out the possibility of an obstacle [1].

Arguments suggestive of acute renal failure will be normal or increased volume (Fig. 9.1). Chronic renal failure would give small kidneys with thinning of the parenchyma and irregular borders (Fig. 9.2). Kidneys can show global dedifferentiation. The parenchyma can resemble the sinus (parenchymocentral dedifferentiation), or, within the parenchyma, medullary pyramids and cortex can be hard to detect (corticomedullary dedifferentiation). However, these patterns do not seem useful in emergency situations.

Acute pyelonephritis is usually barely or not accessible to two-dimensional ultrasound, but severe forms can sometimes be diagnosed. Figure 9.3 shows the routine ultrasound of a 52-year-old female, admitted for severe sepsis, with massive bilateral enlargement of the kidneys, with no differentiation. Diagnosis was hemorrhagic pyonephritis with diffuse purulent areas.

Parenchymatous candidiasis can sometimes be diagnosed (Fig. 9.4). Emphysematous pyelonephritis, a rare finding, gives gas bubbles within the parenchyma. In the case of severe rhabdomyolysis with acute renal failure, we can observe enlarged kidneys with complete dedifferentiation. Renal trauma is presented in Chap. 24.

A renal cyst is a benign finding. In view of its great frequency, we insert a characteristic example (Fig. 9.5) and a case of renal polycystic disease (Fig. 9.6).

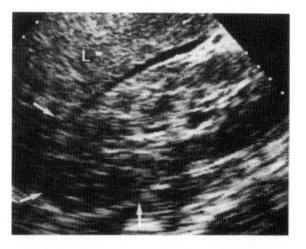

Fig. 9.1. Acute renal failure. The kidney has a homogeneous echoic pattern, i.e., complete dedifferentiation. Kidney and liver (*L*) have the same echogenicity, and the kidney is barely outlined (*arrows*). This scan, as nearly all that follow, is longitudinal

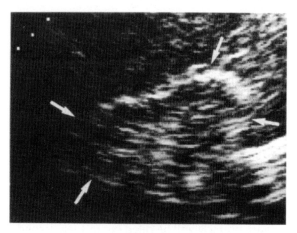

Fig. 9.2. Chronic renal failure. Small size (*arrows*), thinned parenchyma and irregular borders

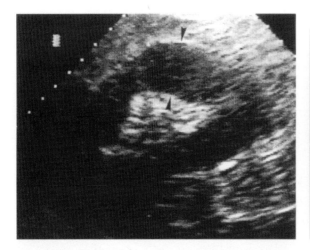

Fig. 9.3. This kidney is frankly enlarged (long axis, 14 cm) and the peripheral area is extremely thickened (*arrowheads*), without differentiation. It was an acute pyonephritis responsible for severe septic shock. Each kidney weighed 500 g and contained multiple areas with pus, necrosis and hemorrhage

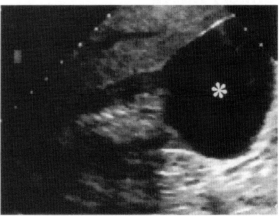

Fig. 9.5. Inferior renal cyst (*asterisk*). The kidney seems to be interrupted, with a ragged edge. This cyst is regular, anechoic. This pattern, here caricatured, should not disconcert since it is regularly observed

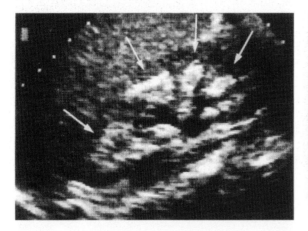

Fig. 9.4. Hyperechoic pattern of the pyramids (*arrows*) in a patient with patent urinary candidosis

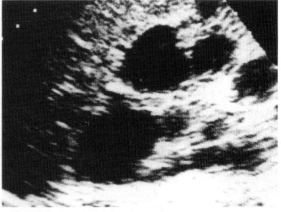

Fig. 9.6. Renal polycystic disease. Cysts have peripheral topography and do not communicate with each other, two features that distinguish it from dilatation of the urinary cavities

Renal Pelvis

Septic shock, increased creatinemia, and a drop in diuresis are daily situations. The possibility of a urinary obstacle will be ruled out in a few seconds if a small ultrasound unit is readily available.

Dilatation of the renal pelvis is rarely but regularly encountered in our experience. Of 400 consecutive critically ill patients, we have had eight cases, 2%. This rate was increased if only sepsis or acute renal failure are considered. Interestingly, the pain is nearly always absent in septic, encephalopathic patients. Causes encountered in the ICU were pelvic hematoma, obstruction of the urinary probe (see Fig. 9.12), bladder distension with overflow, blocked calculi or hydronephrosis (Fig. 9.7) with superimposed pyonephrosis (Fig. 9.8). Detection of dependent echoic patterns within dilated cavities of hydronephrosis is characteristic of pyonephrosis [2]. Pelvic cancer is a classic cause. In trauma, a blood clot can again cause obstructive anuria.

Dilated calices and renal pelvis yield a well-known pattern. The three calices and the pelvis,

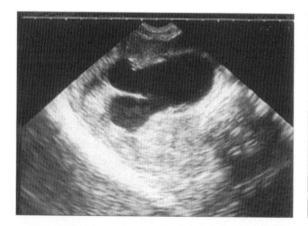

Fig. 9.7. Hydronephrosis. Major dilatation of the renal pelvis. Note the rounded end, which indicates chronic obstruction. This single scan does not show patent signs of acute infection (see Fig. 9.8). Septic shock, transverse scan of the right kidney

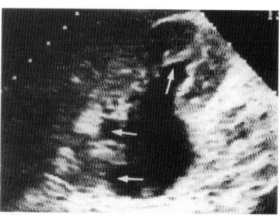

Fig. 9.9. Mild dilatation of the cavities. The pelvis is slightly more dilated. The end of the calyces is concave (*arrows*), usually a sign of acute obstruction. Longitudinal scan, making it possible to visualize the three calyx groups

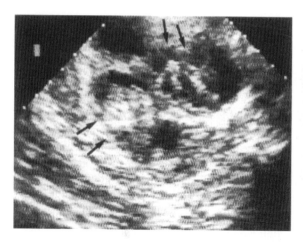

Fig. 9.8. Sequel of Fig. 9.7. The ultrasound scan of the kidney now shows heterogeneous echoic masses within the dilated cavities (*arrows*). These images had a undulating motion in real-time. The diagnosis of superimposed infection (pyonephrosis over hydronephrosis) can now be put forward

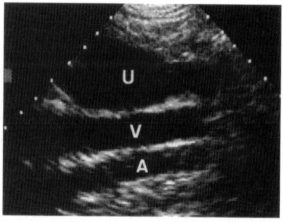

Fig. 9.10. In this rarely obtained longitudinal scan of the right flank, one can observe a dilated ureter (*U*), inferior vena cava (*V*) and abdominal aorta (*A*). The ureter is usually masked by bowel gas

normally virtual or barely visible, are clearly depicted here, anechoic and communicating with each other (Fig. 9.9). In polycystic disease, the numerous cysts do not communicate with each other (see Fig. 9.6). It should be noted that a moderate dilatation, with persistence of the concavity of the calyces, indicates acute obstruction. Conversely, massive dilatation evokes chronic obstruction, with the end of the calices rather bulged and thinned parenchyma (Fig. 9.7).

A dilated ureter is rarely accessible, since intra-abdominal structures are usually in the way (Fig. 9.10).

Dilatations of calices and pelvis without obstruction are possible, and attributable to chronic infectious episodes, rare causes and the ampullary pelvis, a variant of the normal pelvis, which should affect 8% of the population [3]. For some authors, however, this pattern means occult obstruction [4], which should be recognized and treated. In our observations, recognition of a unilateral and moderate dilatation in a septic patient should not be considered fortuitous.

A parapyelitic cyst or hypoechoic fat may, in a hasty examination, simulate renal dilatation.

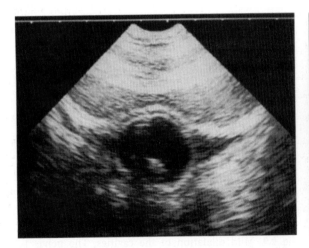 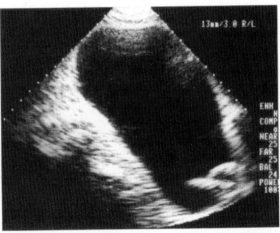

Fig. 9.11. In this suprapubic transversal scan, probe pointing toward the rear and the bottom of the patient, one can see a regular round and medial structure, the inflated balloon of the urinary probe. The bladder is here correctly drained, thus virtual (compare with Fig. 9.12)

Fig. 9.12. Major bladder distension in spite of a urinary probe. The balloon and the end of the probe are visible. The probe was obstructed here. Longitudinal suprapubic scan

An acute obstruction may not yield dilatation of the calices if they have lost their compliance (fibrosis, retroperitoneal malignancy), or sometimes because of major hypovolemia [5–7]. Only iodine explorations would make the diagnosis of obstruction, but we have not yet encountered this situation.

Bladder

This organ is easy to explore and can provide highly contributive information in daily practice. As a rule, a catheterized bladder is empty. A careful examination should, if necessary, identify the balloon of the probe, which seems lost in the pelvis, but always medial (Fig. 9.11).

A bladder that is probed but not empty is not a normal finding (Fig. 9.12). In this case, the bladder should be repeatedly examined in order to check that the trapped volume does not increase. Rarely but regularly, we observe a genuine distended bladder. If this occurs in a sedated patient with circulatory support, the physician can make a wrong diagnosis of anuria and increase drugs or fluid therapy, before the distension becomes clinically obvious. Therefore, if anuria occurs in such patients, the first reflex should be to check if there is no occult bladder retention.

Similarly, in an obese patient, the clinical diagnosis of distended bladder can be difficult. It will always be immediate with ultrasound. Distended bladder is maybe one of the most obvious diagnoses for the beginner in ultrasound (see Fig. 9.12). The ultrasound probe, applied just over the pubis, detects a huge liquid mass which is medial, round, and near the anterior wall. The pitfalls are easily avoidable. The most frequent is the peritoneal effusion that mimics a distended bladder. If a single transversal scan is performed at the pelvis, peritoneal effusion can have a roughly square section and look like a moderately distended bladder (Fig. 9.13). However, peritoneal effusion appears to open when the probe scans more cranially, whereas a bladder appears to close.

In the female, the association of peritoneal effusion and a full bladder will yield a complex but characteristic pattern, the Thai dragoon head sign (Fig. 9.14).

In a recently anuric patient, a daily ultrasound can detect recovery of diuresis. This procedure does not last more than 10 s and should prevent a prolonged and useless insertion of a urinary probe.

The bladder content can be informative. A blood clot yields an echoic dependent pattern. A calculus gives a dependent image with a frank posterior shadow. A purulent retention can have a very characteristic pattern (Fig. 9.15). Last, an enlarged prostate can be detected (Fig. 9.16).

Bladder

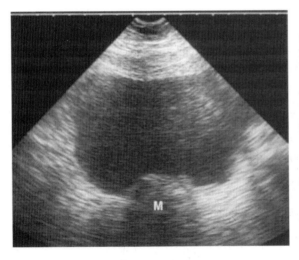

Fig. 9.13. Suprapubic transversal scan. This medial fluid image with square section and a tissular image (*M*) lifting the floor is highly suggestive of a moderately distended bladder. It is in fact peritoneal effusion in the Douglas pouch. The image at *M* is probably a bowel loop. A dynamic scan of the ultrasound probe upward and downward will prevent the error: the bladder will be identified below, and this fluid image will have an opened shape above

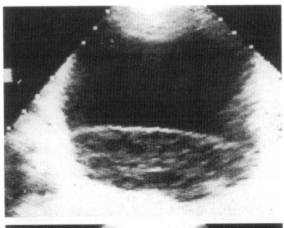

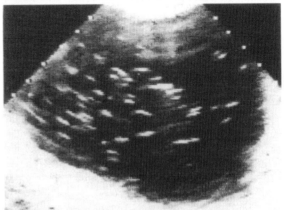

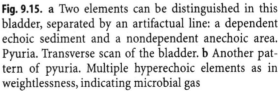

Fig. 9.15. a Two elements can be distinguished in this bladder, separated by an artifactual line: a dependent echoic sediment and a nondependent anechoic area. Pyuria. Transverse scan of the bladder. **b** Another pattern of pyuria. Multiple hyperechoic elements as in weightlessness, indicating microbial gas

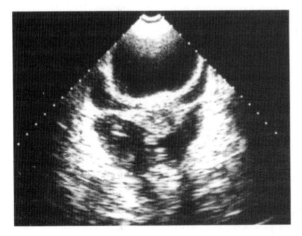

Fig. 9.14. This complex transverse suprapubic scan may intrigue the operator. One can imagine the head of a Thai dragon. This is, in fact, a full bladder associated with peritoneal effusion in a young woman. The bladder is the round shape at the top of the screen. The eyes and the mouth of the dragoon reflect the peritoneal effusion. The nose is formed by the uterus and the large ligaments. The teeth are generated by solid structures floating in the effusion – a hemoperitoneum here

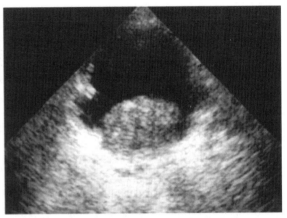

Fig. 9.16. Medial regular tissular mass protruding in the bladder lumen, typical from a prostatic adenoma. This finding is sometimes useful in cases of acute obstructive renal failure

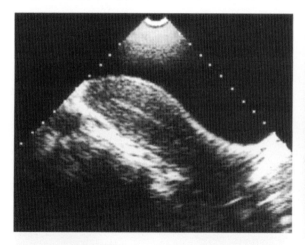

Fig. 9.17. Empty uterus in a long-axis scan, behind the bladder. The vacuity line, which indicates absence of pregnancy, is frankly outlined within the uterine muscle

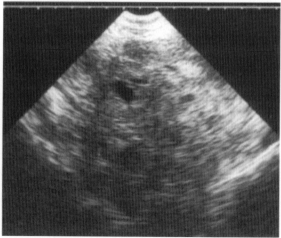

Fig. 9.19. Transverse view of the pelvis in a young female in shock. A motionless echoic mass indicates a massive blood clot in a highly probable ectopic pregnancy. The intensivist should not be asked the precise site of the pregnancy, since the recognition alone of a peritoneal effusion indicates, here, immediate lifesaving surgery

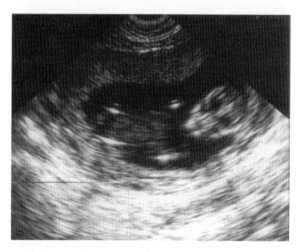

Fig. 9.18. An embryo is visible in the uterus, seemingly observing the viewer. It is like a cat turned on its back, head at the right of the image. This should incite the physician not to overindulge in ionizing radiation procedures

Uterus and Adnexa

We like to take a look at the uterus before any emergency radiological examination, in order to check its vacuity (Fig. 9.17). If a pregnancy is detected (Fig. 9.18), the reader should see Chap. 28, where all details about management are detailed. Occurrence of an acute respiratory disease in a pregnant woman usually raises problems [8]. For this emergency application, we do not need to await full repletion of the bladder. In some postoperative cases where the suprapubic approach is impossible, using a perineal approach is autho-

rized, as is an endovaginal investigation with the small probe covered with a glove. This approach, although not very academic, is perfectly valuable when a distended bladder is sought.

The aim of this book is not to describe gynecological problems such as uterine apoplexy, ectopic pregnancy or others. It suffices to note that pyometritis gives a liquid endouterine image. Hyperechoic punctiform images (gas) are a strong argument if there is severe pelvic sepsis [9]. Endometritis produces diffuse swallowing of the parenchyma [9]. Ectopic pregnancy shows a subtle direct image for the specialist, and a rough indirect image for the nonspecialist, the hemoperitoneum. Note that this hemoperitoneum can be echoic at the first examination, thus particularly misleading (Fig. 9.19). The syndromes of defibrination can provide information on a rapid ultrasound confirmation, although history and physical data are generally sufficient for the decision of a lifesaving hysterectomy.

Renal Transplantation

A grafted kidney usually lies in the fossa iliaca. Surgical complications are more accessible to ultrasound than medical complications. Postoperative collection can be caused by abscess, hematoma, lymphocele or urinoma. They can be explored with an ultrasound-guided tap. Dilatation of the calices suggests obstruction caused by edema or anastomotic stenosis of the ureter. Stenosis of the renal artery should be explored with Doppler.

Medical complications, in spite of numerous signs of acute or chronic rejection, cyclosporine toxicity or tubulointerstitial nephritis, are generally diagnosed by renal biopsy [10].

Interventional Ultrasound

Percutaneous nephrostomy makes it possible to treat a urinary obstruction and to drain infected urine at the bedside if ultrasound-guided. The kidney is punctured by the posterior or posterolateral approach. The colon and the pleura are thus avoided. A needle is inserted in the dilated cavities. Urine is collected for analysis. A guide is then introduced through the needle. A drainage catheter is inserted, sometimes after several dilatations.

We have had the opportunity to use this technique to treat a patient who could not be moved as she had severe septic shock. Later, procedures were performed in the radiology department for localizing the obstruction level and in the operating room for removing the calculus, at this time in a stabilized patient.

Percutaneous nephrostomy is a procedure whose mortality rate (0.2%) is said to be lower than that of surgery [3]. Complications include hemorrhage or infection and should be balanced with the advantages.

If suprapubic catheterization is indicated, ultrasound guidance provides visual monitoring. A penetration site more cranial than classically done should theoretically limit the risk of sepsis of the prevesical space. Digestive interpositions can then be ruled out using ultrasound.

References

1. Resnick MJ and Rifkin MD (1991) Ultrasonography of the urinary tract. Williams and Wilkins, Baltimore
2. Subramanyan BR, Raghavendra BN, Bosniak MA et al (1983) Sonography of pyelonephrosis: a prospective study. Am J Roentgenol 140:991–993
3. Finas B, Mercatello A, Tognet E, Bret M, Yatim A, Pinet A, Moskovtchenko JF (1991) Stratégies d'explorations radiologiques dans l'insuffisance rénale aiguë. In: Goulon M (ed) Réanimation et Médecine d'Urgence. Expansion Scientifique Française, Paris, pp 153–174
4. Laval-Jeantet M (1991) La détection de maladies graves par échographie systématique chez le généraliste. Presse Med 20:979–980
5. Goldfarb CR, Onseng F, Chokshi V (1987) Nondilated obstructive uropathy. Radiology 162:879
6. Maillet PJ, Pelle-Francoz D, Laville M, Gay F, Pinet A (1986) Nondilated obstructive acute renal failure: diagnostic procedures and therapeutic management. Radiology 160:659–662
7. Charasse C, Camus C, Darnault P, Guille F, Le Tulzo Y, Zimbacca F, Thomas R (1991) Acute nondilated anuric obstructive nephropathy on echography: difficult diagnosis in the intensive care unit. Intensive Care Med 17:387–391
8. Felten ML, Mercier FJ, Benhamou D (1999) Development of acute and chronic respiratory diseases during pregnancy. Rev Pneumol Clin 55:325–334
9. Ardaens Y, Guérin B, Coquel Ph (1990) Echographie pelvienne en gynécologie. Masson, Paris
10. Cauquil P, Hiesse C, Say C, Vardier JP, Cauquil M, Brunet AM, Galindo R, Tessier JP (1989) Imagerie de la transplantation rénale. Feuillets de Radiologie 29:469–480

The Retroperitoneal Space

The kidneys have been described in Chap. 9.

Abdominal Aorta

Abdominal aortic analysis should be routine in any critical situation. The examination should be done gently, in order to avoid any uncontrolled pressure. Bowel gas can be a source of failure. However, a left translumbar approach can bypass the anterior gas obstacles.

Basic signs of abdominal aortic aneurysm are a loss of parallelism of the aorta walls with a fusiform or sometimes sacciform shape (Fig. 10.1). If local conditions are favorable, ultrasound will provide, like CT, a global overview of the lumen, thrombosis, wall thickness (increased in the case of inflammation) and collateral vessels. In the case of leakage, a collection will be found in the left retroperitoneal space (Fig. 10.2). In one rare case, it was possible to observe a precise area of whirling in rhythm with heart frequency, within the hematic effusion. This dynamic pattern obviously indicated the location of the leakage. This observation was serendipitous, and indicated extremely urgent surgery.

Fortuitous discovery of incipient aneurysm is frequent in the medical ICU and should prompt further investigations. An atherosclerotic aorta with irregular borders is a sign indicating that the patient may have diffuse potential arterial damage.

A dissection of the abdominal aorta yields enlarged lumen with an intimal flap separating two channels. When the aorta can be followed to its bifurcation, the progressive disappearance of one channel can be noted (Fig. 10.3).

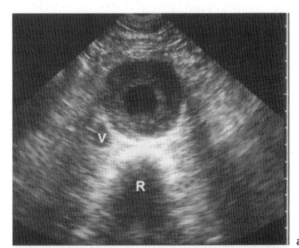

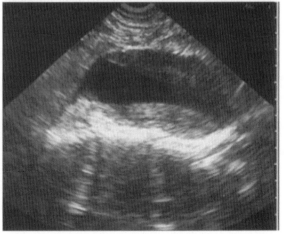

Fig. 10.1. a Transverse scan of the epigastric area. The aorta is recognized by its location anterior to the rachis (*R*), at the left of the inferior vena cava (*V*). A substantial enlargement of its caliper is immediately noted. There is a large thrombosis within the aneurysm, with a tissue-like peripheral layer and quasi-normal caliper of the lumen. A simple aortography would obviously underestimate this aneurysm. **b** Longitudinal scan, specifying the extension of the aneurysm

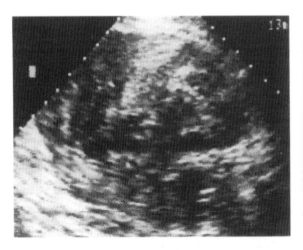

Fig. 10.2. Patient in shock with abdominal pain. Huge echoic heterogeneous roughly rounded mass with anterior contact (transversal scan, left flank approach). Acute retroperitoneal hematoma, with early clotting

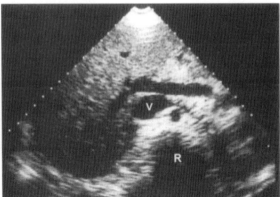

Fig. 10.4. The caliper of the abdominal aorta in this young female in shock appears extremely low (9 mm). It may correspond to major vasoconstriction or hypovolemia. Epigastric transverse scan. *V*, inferior vena cava; *R*, rachis

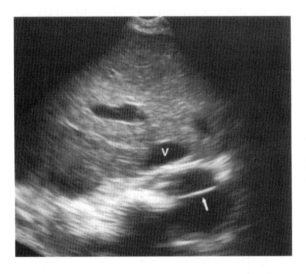

Fig. 10.3. Epigastric transversal view in a patient in shock with thoracoabdominal pain. Throughout the liver and at the left of the inferior vena cava (*V*), the abdominal aorta is clearly visible. It is possible to detect a flap (*arrow*, which was positioned at the level of the true channel) separating the aortic lumen into two parts. When the probe moves downward, the superior channel (false channel) progressively vanishes

Other Information Available from Abdominal Aorta Study

For maximal use of the full potential of noninvasive ultrasound, it may be of interest to investigate the aortic caliper in patients in shock.

One hypothesis is that this caliper diminishes when there is vasoconstriction. We know that in case of vasoparalysis, only arteriolar resistances can be altered. However, early findings indicate that the large-vessel caliper can also be variable (Fig. 10.4). The aorta should be supple, not atheromatous. The measurement is taken at a precise and therefore reproducible level. We propose crossing with the left renal vein.

Retroperitoneal Hematoma and Other Disorders

Ultrasound finds a generally voluminous mass, heterogeneous, with often a dependent zone that is rather echoic, corresponding to blood clots, and a nondependent area that is rather poorly echoic, corresponding to the serum. This area can be rich in septations due to fibrin deposits (Fig. 10.2). It is possible to follow this hematoma up to the insertion of the psoas muscle. However, we must admit that subtle signs are rarely required in often plethoric patients. Peritoneal blood effusion can be associated with contiguity and should not be misleading.

A posterior translumbar approach is logical, but an extensive hematoma generally comes in contact with the anterior abdominal wall (clinically detectable). The differential diagnosis with a parietal hematoma, whose treatment is different, will be resolved by studying the linking angles.

When a superinfection is suspected, an exploratory, ultrasound-guided tap is possible.

Pneumoretroperitoneum should theoretically yield a characteristic image, since air stops the ultrasound beam.

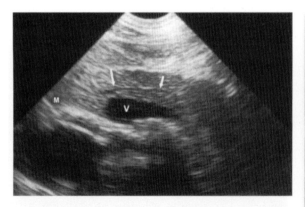

Fig. 10.5. In this transverse epigastric scan, the pancreatic parenchyma is perfectly identified, homogeneous, with a well-defined main pancreatic duct (*arrows*), end of the common bile duct (*M*) and confluence of the portal and mesenteric superior veins (*V*). Normal pancreas

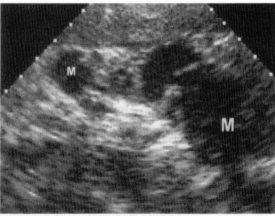

Fig. 10.7. Hemorrhagic necrotizing acute pancreatitis, transverse scan. The pancreas can be identified only using the vascular landmarks. Numerous hypoechoic collections along the head (*m*) and the body (*M*)

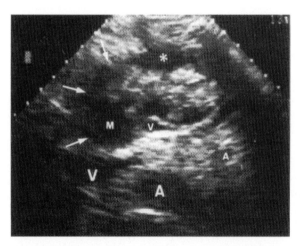

Fig. 10.6. Hemorrhagic necrotizing acute pancreatitis. The head and body of the pancreas are enlarged (*arrows*) and heterogeneous. A hypoechoic image can be distinguished within the head (*M*), and a collection surrounding the pancreatic space, anterior to the body (*asterisk*). A, aorta; a, superior mesenteric artery; V, inferior vena cava, v, splenic vein. Transverse scan

Inferior Vena Cava

The inferior vena cava is studied in Chap. 13.

Pancreas

Precisely localized using the vascular landmarks (see Fig. 4.6, p 21), the pancreas can be hard to detect since there is a frequent reflex ileus [1]. However, gas collections can be mobilized, and the stomach can be filled with liquid in order to create an acoustic window. In favorable cases, the study is contributive, and the main pancreatic duct and all the bile ducts can be studied (Fig. 10.5). Maximal dimensions of a normal pancreas are 35 mm at the head, 25 mm at the isthmus and 30 mm at the body [2].

Acute pancreatitis is a familiar field in radiology [3]. The organ has increased in size, with a hypoechoic heterogeneous pattern. Necrotic roads can be observed in the pancreatic space (Fig. 10.6) but are also very remote. In some instances, the pancreas can have a normal pattern [4].

CT is usually indicated in first-line investigations for the positive diagnosis of acute pancreatitis, since gases are not a hindrance, and a regional and remote analysis is easy to do. Ultrasound is used for monitoring after an initial CT. Iterative ultrasound scans detect the appearance of fluid within the pancreas, surrounding it, or from a distance. Venous thrombosis (splenic or superior mesenteric veins) is accessible (see p 38, Chap. 6). The constitution of false aneurysms (mainly the superior mesenteric artery) can be monitored.

The appearance of a collection (whose echogenicity can be variable) can be caused by simple necrosis or infectious abscess (Fig. 10.7). Ultrasound can answer the question by tapping the collection, provided there is no bowel or vascular interposition. One disorder must be ruled out before any tap: false aneurysm. Doppler is usually able to answer this question, but if two-dimensional ultrasound identifies dynamic changes within

A pancreatic pseudocyst produces a well-defined, anechoic image with a thin regular wall. The size is often substantial. Dependent echoes suggest superinfection.

Vertebral Disks

The rachis, which is the posterior limit of the retroperitoneum, stops the ultrasound beam. However, ultrasound can go through intervertebral disks. It is then possible to analyze unusual structures such as the content of the spinal canal (Fig. 10.8). We have not given this analysis a particular relevance (should meningitis yield a particular pattern?), but Fig. 10.8 is a striking example of the still untapped features of ultrasound.

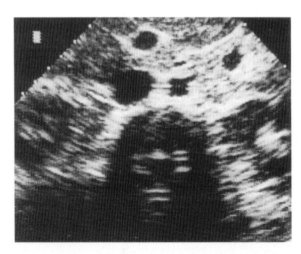

Fig. 10.8. This ghostly apparition seemingly observing the viewer, here intended to relax the reader, shows how well ultrasound can perform. In this transverse scan passing through an intervertebral disk, the spinal canal and the intervertebral foramen are well defined, forming the nose and eyes of the creature. Depending on one's imagination, a gorilla in the mist or one of the main characters from the »Star Wars« movies may become visible

the collection, it can also answer the question: slow, nonsystematized particle movements can be safely tapped. Whirling systolic movements, when visible, clearly indicate false aneurysm. An exploratory tap with thin material is easy and distinguishes abscess from necrosis. An evacuation procedure requires large, invasive material since the collection can contain large debris. Some authors recommend surgery for central collections, and percutaneous procedures for peripheral ones [5].

References

1. Silverstein W, Isckoff MB, Hill MC, Barkin J (1981) Diagnostic imaging of acute pancreatitis: prospective study using computed tomography and sonography. Am J Roentgenol 137:497
2. Weill FS (1985) Pathologie pancréatique. In: Weill FS (ed) L'ultrasonographie en pathologie digestive. Vigot, Paris, pp 345–375
3. Freeny P, Lawson TL (1982) Imaging of the pancreas. Springer Verlag, Berlin Heidelberg New York
4. Lawson TL (1978) Sensitivity of pancreatic ultrasonography in the detection of pancreatic disease. Radiology 128:733
5. Lee MJ, Rattner DW, Legemate DA, Saini S, Dawson SL, Hahn PF, Warshaw AL, Mueller PR (1992) Acute complicated pancreatitis: redefining the role of interventional radiology. Radiology 183:171–174

Spleen, Adrenals, and Lymph Nodes

These disparate organs are artificially collected in a single chapter.

Spleen

Ultrasound can diagnose splenomegaly. The probe must be applied rather posteriorly at the last intercostal spaces. In a supine patient, the distal part of the probe should in practice sink into the bed. A normal spleen can be difficult to see, since it can be surrounded by lung air and bowel gas. Conversely, an enlarged spleen is easily diagnosed. What is more, the homogeneous or heterogeneous pattern of the parenchyma can be appraised (Fig. 11.1). In an obese patient, for instance, ultrasound will be of precious help, even if some think the diagnosis of splenomegaly remains clinical.

Splenomegaly can also create an acoustic window making the analysis of the following organs accessible: adrenals, kidney, tail of the pancreas, stomach, and aorta.

Splenic abscess in the critically ill is often occult, with a paucity of clinical signs. In the minimal cases, ultrasound can be normal, showing only an apparently homogeneous enlarged spleen, whereas CT shows the abscess perfectly (Fig. 11.2). In intermediate cases, the abscess is isoechoic to the spleen, but is separated from the normal parenchyma by a thin dark border that clearly outlines the pathological mass (Fig. 11.3). Usually, abscesses yield hypoechoic heterogeneous images (Fig. 11.4). Hemorrhagic splenic suppuration accompanying stercoral peritonitis can yield hypoechoic enlarged spleen with liquid-like areas and hyperechoic elements caused by microbial gas (Fig. 11.5). Last, the spleen can be discretely heterogeneous, not to say normal, in genuine fulminant tuberculous miliaries (Fig. 11.6).

Perisplenic effusion (see Fig. 5.3), a traumatic rupture of the spleen (irregular intraparenchymatous image, with capsular hematoma), and a

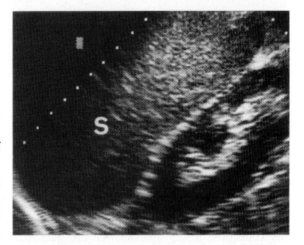

Fig. 11.1. Splenomegaly (*S*) covering the entire left kidney. This homogeneous spleen is 16 cm long. Longitudinal scan of the left hypochondrium

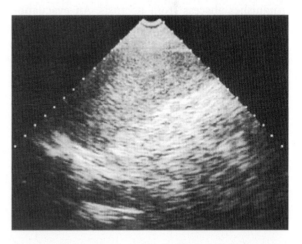

Fig. 11.2. This spleen was considered homogeneous using ultrasound, whereas CT revealed an abscess. In these cases, especially in plethoric, poorly echoic patients, the poor echogenicity of the image should be recognized, in order to request other imaging modalities

Interventional Ultrasonography of the Spleen

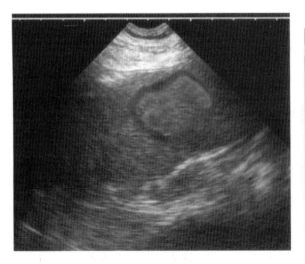

Fig. 11.3. Splenic abscess isoechoic to the spleen. However, a thin stripe is noted. Septic shock in a 68-year-old female who had had cold abdominal surgery 1 month before, and without focal clinical signs at the time of the examination

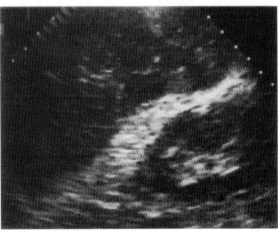

Fig. 11.5. Hypoechoic and heterogeneous splenomegaly in a septic patient. Surgery revealed stercoral peritonitis with hemorrhagic suppuration of the spleen

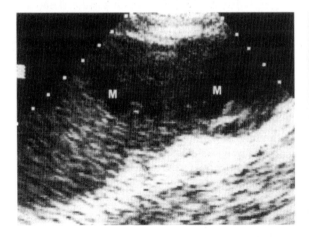

Fig. 11.4. Hypoechoic images (*M*) within an enlarged spleen. The tap revealed pus with staphylococcus. Splenic abscesses complicating endocarditis in a 48-year-old male

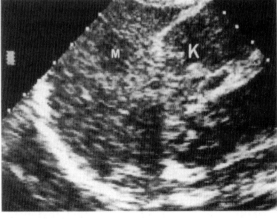

Fig. 11.6. This spleen has normal dimensions and quasi-normal echostructure, except for some mildly hypoechoic areas (*M*). Autopsy of this young man with septic shock revealed diffuse tuberculous miliary, including the spleen. The mildly granulose pattern of the spleen was slightly questionable when subsequently reading the examination. Longitudinal scan. *K*, left kidney

splenic infarct (regular pyramidal hypoechoic image) can also be diagnosed (Fig. 11.7). Splenic infarct can become superinfected. Homogeneous splenomegaly is common in portal hypertension. On occasion, splenic artery aneurysm can be recognized. More relevant in daily practice is the possibility of locating the spleen before any left thoracentesis (see Fig. 15.7, p 99).

Interventional Ultrasonography of the Spleen

The spleen, a peripheral organ, is a possible target for interventional procedures. Percutaneous drainage of splenic abscesses is an alternative to surgery [1–3]. Described complications are hemorrhage or infections, but, although spontaneous mortality of splenic abscess is 100% and 7.8% if surgically treated [4], it is only 2.4% after percutaneous procedures [3].

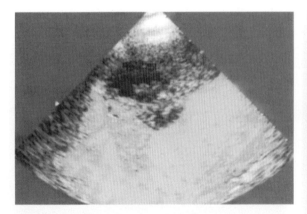

Fig. 11.7. Splenic infarction. Roughly pyramidal hypoechoic image with peripheral base

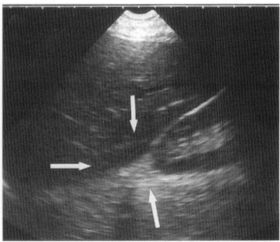

Fig. 11.9. If not detecting the adrenal itself, ultrasound can expose the adrenal space perfectly, here between liver and right kidney. This area is currently being investigated in our septic patients

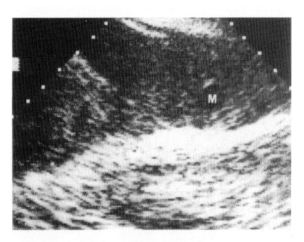

Fig. 11.8. This figure is the sequel to Fig. 11.4, after evacuation of the abscess. The target is significantly reduced

Some authors propose a simple therapeutic aspiration with a 18- to 19.5-gauge needle as a first line of treatment. Antibiotics can possibly be injected in situ [3]. With a 21-gauge needle, we have diagnosed staphylococcus abscess (see Fig. 11.4) and subsequently aspirated it (Fig. 11.8), without hemorrhagic or infectious complications.

Adrenals

Imaging the adrenals in emergency situations is without doubt of limited value. However, the potential impact of corticotherapy in septic shock can be a reason for new interest in this exploration if it is accepted that accurate detection of adrenal necrosis will alter management. It is assumed that CT will be more accurate than ultrasound, but this requires transportation of a very unstable patient.

The adrenals are usually not visible. They are surrounded by fat covering the kidney (Fig. 11.9). Ultrasound signs of the acute adrenals have been described insufficiently in the literature. In the case of bilateral hemorrhagic necrosis, an echoic mass over the kidney has been described [5, 6]. Pheochromocytoma can sometimes yield a voluminous mass. Other conceivable applications, although of limited clinical value, will be the search for an adrenal tumor in a patient admitted for severe arterial hypertension, for adrenal metastases, and last, assessment of acute adrenal failure.

Enlarged Lymph Nodes

Voluminous lymph nodes can create obstructions, for instance of the bile ducts. The diagnosis is based on one or several masses, round or egg-shaped, tissular, and above all located along the vascular axes (see Fig. 12.8, p 73). Detection of lymph node enlargement allows making certain diagnoses but, without exception, the definitive exploration will be made after the critical period.

References

1. Berkman WA, Harris SA Jr, Bernardino ME (1983) Non-surgical drainage of splenic abscess. Am J Roentgenol 141:395–397
2. Lerner RM, Spataro RF (1984) Splenic abscess: percutaneous drainage. Radiology 153:643–647

3. Schwerk WB, Görg C, Görg K, Restrepo I (1994) Ultrasound-guided percutaneous drainage of pyogenic splenic abscesses. J Clin Ultrasound 22:161–166
4. Nelken N, Ignatius J, Skinner M, Christensen N (1987) Changing clinical spectrum of splenic abscess: a multicenter study and review of the literature. Am J Surg 154:27–34
5. Enriquez G, Lucaya J, Dominguez P, Aso C (1990) Sonographic diagnosis of adrenal hemorrhage in patients with fulminant meningococcal septicemia. Acta Paediatr Scand 79:1255–1258
6. Mittelstaedt CA, Volberg FM, Merten DF, Brill PW (1979) The sonographic diagnosis of neonatal adrenal hemorrhage. Radiology 131:453–457

CHAPTER 12

Upper Extremity Central Veins

Nearly all of the central venous axes are accessible to ultrasound (Fig. 12.1). The applications are numerous in the critically ill:

- Recognizing small or even collapsed veins, the very ones whose catheterization will be problematic
- Diagnosing venous thrombosis occurring on the central catheter
- Recognizing the correct position of a central venous catheter
- Estimating blood volume in a patient in shock (see Chap. 13)
- Assisting in rapid central venous catheterization, a sometimes thorny situation in the emergency context

We will first discuss the internal jugular vein, which is familiar to the intensivist, and will then see in detail the subclavian vein, which we prefer for central venous access.

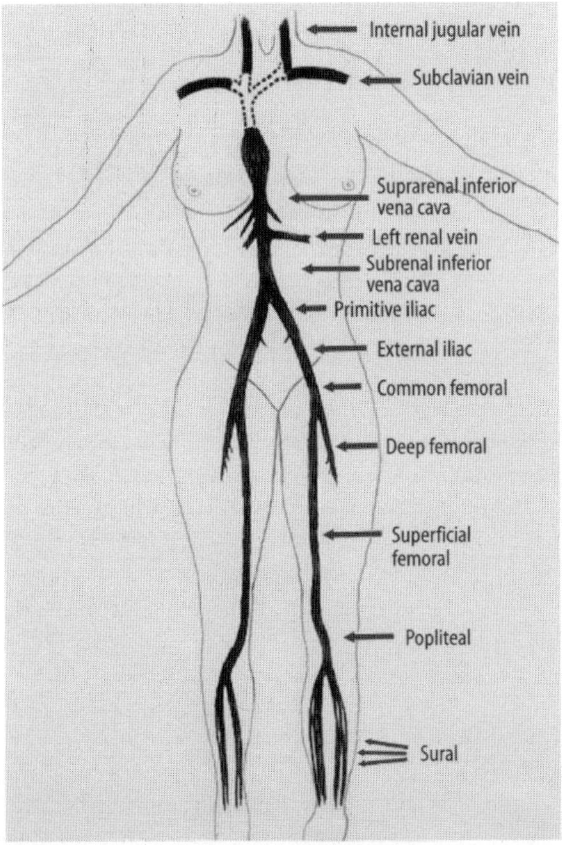

Internal Jugular Vein: Normal Pattern

The internal jugular vein is recognized outside the carotid artery (Figs. 12.2, 12.3). The carotid artery is small, with a perfectly round cross-section. The cross-sectional shape of the vein can be round, oval, triangular or even collapsed. In the longitudinal approach, the borders of the vein are never perfectly parallel (although in the artery they are). Dynamic changes in the vein are often vast, whereas the artery has small, halting systolic expansion. More than ever, the probe should be held like a

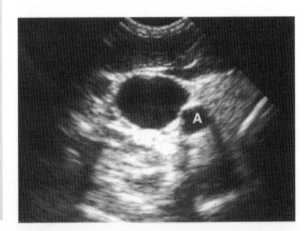

Fig. 12.1. This figure shows the deep venous axes that are accessible to ultrasound. The vena cava superior and the brachiocephalica vein, often hard to detect, are indicated with dotted lines

Fig. 12.2. Normal right internal jugular vein, transverse scan. The vein is located outside the artery (*A*), has a round shape, a caliper of 13×20 mm, and an anechoic lumen

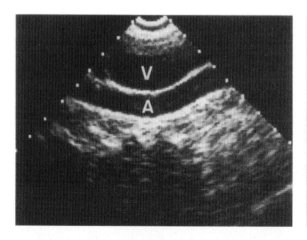

Fig. 12.3. Normal internal jugular vein (*V*) in a longitudinal scan. In this scan, the vein lies anterior to the artery (*A*), not a rare finding

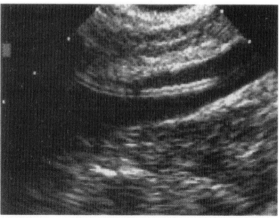

Fig. 12.4. A catheter is clearly identified within the venous lumen. This pattern (two strictly parallel hyperechoic lines) is characteristic of any catheter. The route through the soft tissues is also visible here

pen, as even a slight pressure (not to mention the weight of the probe alone) can contribute to collapsing the vein.

The ultrasound analysis of this vein, as any other vein, can be separated into three steps:

1. Static analysis in static approach: the probe is applied over the vein, the operator simply observes. It is already possible to study the venous area, search for asymmetric caliper between the right and left vein (see »Ultrasound and Central Venous Catheterization«), confirm the presence of a catheter (Fig. 12.4) and detect thrombosis. The internal jugular vein is different from the other veins because it is always highly accessible to ultrasound and the parasite echoes are generally absent. Using these features, an anechoic pattern is a basic sign of absence of thrombosis.
2. Dynamic analysis in the static approach: this step provides information on the variations in venous caliper, as well as the behavior of a venous thrombosis in the lumen. Any central vein has respiratory changes: inspiratory flattening (up to the collapse) in spontaneous ventilation, or conversely inspiratory enlargement in mechanical ventilation. Heart-rhythm changes can also occur.
3. Dynamic approach: the probe's pressure as applied by the operator's hand checks for venous patency (Fig. 12.5). It should be emphasized that this maneuver is not insignificant. A reasonable pressure should be applied. The reasonable limit not to exceed, according to our experience, is between 0.5 and 1 kg/cm^2. This basic notion will be recalled in Chap. 14. If venous thrombosis has previously been recognized using static analysis, any compression technique will be redundant, and could therefore be potentially dangerous.

Internal Jugular Vein Thrombosis

The possibility of diagnosing internal jugular vein thrombosis with ultrasound has been described [1]. In our experience, this is a highly accessible field, using a very simple and feasible technique (Fig. 12.6).

The venous lumen is no longer anechoic. It contains an irregular echoic mass. This pattern has remained constant in our observations. With increasing experience, a compression maneuver is superfluous in typical cases, which are the great majority: the ultrasound pattern is characteristic.

The main pitfall will worry only beginners: ghost echoes generated by nearby hyperechoic surroundings (Fig. 12.7). These parasite echoes never have the anatomical pattern of a genuine thrombosis. They have the features of any artifact: geometrical disposition in the screen, either parallel or meridian. Among other possibly confounding factors, we can cite the sternocleidomastoideus muscle (a pitfall than can easily be avoided) and the enlarged lymph node. Both can simulate venous thromboses, but a simple scan immediately shows that the suspected structures are not

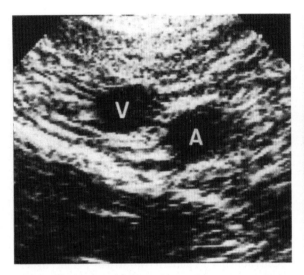
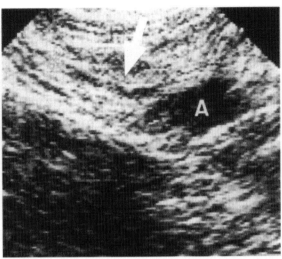

Fig. 12.5. Any vein must normally collapse completely when pressure is exerted by a probe, and this is the case in the figure on the *right* (*arrowhead*). Transverse scan of the subclavian vein (*V*), with the satellite artery (*A*)

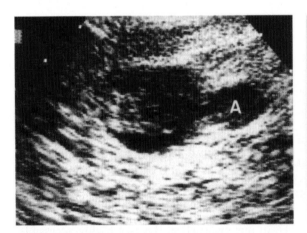
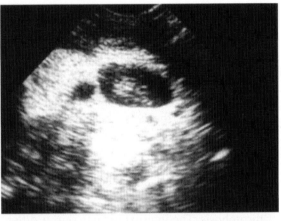

Fig. 12.6. On this transverse scan of the cervical vessels (*A*, artery), the venous lumen is filled by an echoic thrombosis. This thrombosis has homogeneous pattern and is subocclusive: the free lumen is reduced to an anechoic moon shape (which should disappear when applying mild pressure)

Fig. 12.7. This echoic image, in the lumen of the left internal jugular vein, has regular shape. Mild pressure of the probe completely collapses the lumen. This is a ghost artifact. Left carotid artery at the *left* of the vein

tubular. A lymph node has a beginning and an end, whereas a vessel has no end (Fig. 12.8). In some cases, a very tense patient can contract the cervical muscles, and the vein can then be hard to compress.

Venous thrombosis can have multiple characteristics.

Occlusive Thrombosis

Thrombosis can be totally (Fig. 12.9) or partially (Fig. 12.6) occlusive. A controlled compression maneuver does not alter the area of a totally thrombosed vein. Another sign can be called the flight sign and is sometimes useful: the compression maneuver creates a slight movement of the entire vein. In a normal subject, the compression drives back the proximal wall toward a distal wall, which remains in place. If there is a flight sign, the compression maneuver should immediately be interrupted: occlusive venous thrombosis is definitely diagnosed. In a partial occlusion, the compression maneuver easily collapses the free lumen. It should be emphasized that moderate pressure of the probe is necessary and sufficient to collapse a normal vein. The same moderate pressure is enough to

Internal Jugular Vein Thrombosis

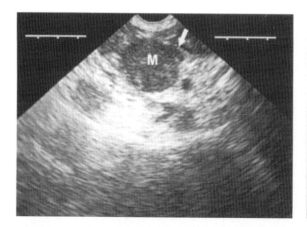

Fig. 12.8. Transverse scan of the neck. A tissue-like mass (*M*) is detected outside the artery. This is an enlarged lymph node, an egg-shaped structure when scanned. In a single scan, venous thrombosis may have the same pattern. The *arrow* designates the shifted and flattened internal jugular vein

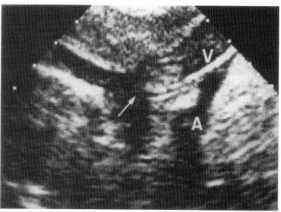

Fig. 12.10. Internal jugular vein (*V*) in a longitudinal axis, probe applied in the supraclavicular fossa. A thrombosis is detected. The *arrow* designates its caudal end, just at the Pirogoff confluent. In real-time, this thrombosis has worrying halting dynamics in rhythm with the heart cycle. Negative progression. *A*, arterial vessels

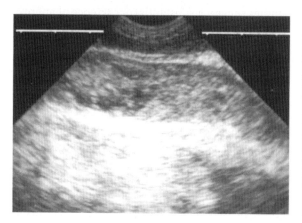

Fig. 12.9. Completely occlusive thrombosis of the left internal jugular vein in a long axis. We can measure at least 6 cm of extension

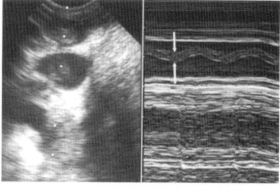

Fig. 12.11. Floating thrombosis of the internal jugular vein. *Left*, real time. *Right*, time-motion. Flagrant dynamics of the thrombosis are objectified: characteristic undulating pattern (*arrows*)

assert absence of compressibility of the vein. Any additional pressure causes an unstable thrombosis to migrate.

Recognition of a complete occlusion obviously renders any catheterization futile.

Long-Axis Extension

The thrombosis can be localized or spread out in the venous long-axis (Fig. 12.9). Its end can sometimes be visualized (Fig. 12.10).

Dynamic Pattern in the Static Approach: Floating Thrombosis

The thrombosis can be motionless. It can also appear to be floating in the static approach. The real-time highly characteristic pattern of a floating thrombosis should make ultrasound the gold standard (Fig. 12.11). Sometimes, characteristic halting movements of the thrombus are observed, in rhythm with the heart cycle: the floating thrombus seems to be attracted by the right heart. With a little imagination, one can guess the short-term future of such thromboses.

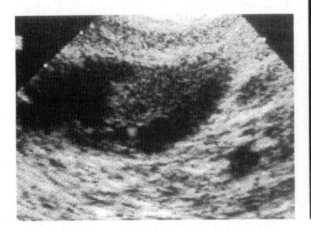

Fig. 12.12. Diaphanous curls are freely floating in the lumen of the internal jugular vein. Since a part of this image is fixed against the wall, one cannot evoke simple echoic flow with visible particles. This pattern is possibly the first step of a rising venous thrombosis

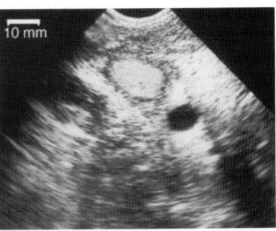

Fig. 12.13. Complete thrombosis of the right internal jugular vein, transverse scan. A thickened wall and an extremely echoic pattern are unusual. Suppurative thrombophlebitis

Dynamic Pattern in the Dynamic Approach

The thrombosis can be somewhat soft or rigid. A recent thrombosis is soft and can be flattened by probe pressure [2], but we hesitate to compress the thrombosis too aggressively in such cases. We have observed the birth of an internal jugular thrombosis, using iterative examinations. First we observed a kind of diaphanous image within the venous lumen. It was more or less fixed against the wall, freely floating in the lumen (Fig. 12.12). At this step, the vein was totally compressible. Twenty-four hours later, a complete, patent thrombosis was present.

Echogenicity

The thrombosis is most often moderately echoic. In cases of infected thrombophlebitis, the pattern can be frankly hyperechoic (Fig. 12.13). Hyperechoic punctiform echoes usually indicate infectious gas bubbles, i.e., septic thrombophlebitis. A thickened wall should also theoretically be observed.

Origin and Outcome

Routine examination of critically ill, ventilated patients shows a high rate of internal jugular thrombosis. A recent catheterization is almost always the explanation (see Fig. 12.18), but some cases occur without any local procedure. Rare studies suggest a substantial occurrence of approximately 70% [3, 4]. The consequences of such studies are usually neglected. Few studies have evaluated these consequences in terms of pulmonary embolism as well as in septic disorders [3]. Pulmonary embolism from upper extremity veins is estimated at 10%–12% of cases [5, 6]. We have seen several cases of pulmonary embolism where internal jugular or subclavian thrombosis was obviously the origin. A study in progress seems to show that mortality is greatly increased in patients with such thromboses, but we must first eliminate selection bias (thrombosis may occur in the most severely ill patients, etc.). Given that death is a daily occurrence in the ICU, with 20%–30% of patients admitted dying, all possible aggravating factors such as upper extremity venous thrombosis, whose treatment is not at all codified, should be carefully scrutinized, since it can vary from therapeutic abstention to fibrinolysis.

We hypothesize that jugular thrombosis can definitely be responsible for pulmonary embolism. These emboli are possibly small and generally without immediate dramatic consequence. However, if they occur repeatedly, they will be responsible for difficult weaning, unclear and transient dysadaptation episodes, subacute fatigue of the patient or even nosocomial pneumonia. All these elements can prolong the patient's stay in the ICU, with unforeseeable consequences.

Many troubling questions can be raised. When a catheter is entirely covered with thrombosis and then removed, where has the thrombosis gone? When mechanical ventilation is replaced by spontaneous breathing, intrathoracic pressure suddenly becomes negative. What happens to a thrombosis that until now had been very flimsily attached to the wall? When thrombosis is generated by a catheter near the skin, i.e., when there is free communication between skin bacteria and the circulatory system, should such thromboses be systematically considered infected? In consequence, migration of such thrombi may bring the bacteria up to the lungs, resulting in positive lung samples. All these issues will be hard to prove. A large study supported by a particularly open ethical committee and comparing the mortality of a group of patients with systematic full-dose heparin (the classic treatment for all venous thromboses) with a untreated group could reach conclusions on the most adapted management.

In practice, and according to the precaution principle, we avoid the internal jugular approach and insert only subclavian catheters, with ultrasound guidance, since the subclavian route is reputed to be less exposed to infections. Prolonged observation may show a better outcome for these patients.

If a septic thrombophlebitis is suspected, ultrasound-guided aspiration of the thrombus, with bacteriological analysis, can be envisaged at the internal jugular level [7]. Right or wrong, we have not investigated this particular situation to date.

Subclavian Vein, Normal Pattern

The discussion in the previous section should theoretically render this vein more attractive. The subclavian vein is localized in a longitudinal subclavian approach, visible on the transverse scan of the subclavian vessels. The vein differs from the artery (Fig. 12.14) in many details: like the internal jugular vein, it does not have a perfectly round section, nor perfectly parallel walls, but wide movements, possible inspiratory collapse. Valves can be observed here. An echoic flow with visible echoic particles is again sometimes seen. Differentiating the vein from the artery using their respective location is more hazardous since the vessels cross each other. Very near the vein (too near for some) is the lung: a hyperechoic horizontal structure with a dynamic pattern and followed by air artifacts (see Chaps. 15–18).

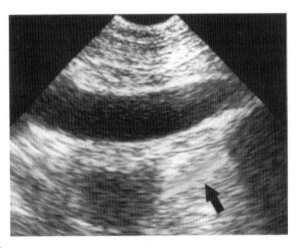

Fig. 12.14. Right subclavian vein in its long axis (transverse scan of the thorax). This vein is free and has a large caliper, favorable to catheterization. Note that the lung surface (*arrow*) is not far

Above all, a nonthrombosed vein can be collapsed by probe pressure (Fig. 12.5), provided the vein is sandwiched between the probe, under the clavicle, and the free hand of the operator above it. The nearer the sternum, the more difficult this maneuver is. For instance, the proximal end of the subclavian veins cannot be compressed. Doppler may be of help here. However, two-dimensional ultrasound is never at a loss for solutions: when a valvule is visualized in this noncompressible area, one should observe its spontaneous dynamics. Frank movements of this valvule (the free subclavian valvule sign) will obviously indicate patency of the vein and will obviate the need for Doppler.

Subclavian Venous Thrombosis

As at the internal jugular level, subclavian thrombosis can be easily identified using the static approach alone. However, spontaneous echogenicity at this level is inferior to that of the cervical area (Fig. 12.15). A mild compression maneuver will show the absence of venous compressibility. The features of internal jugular thrombosis are found at the subclavian area. However, the frequency of subclavian venous thromboses appears strikingly lower than internal jugular thromboses. Except insufficiency of the method, which is improbable, we very rarely encounter patent subclavian venous thromboses after local catheterization.

Subclavian thrombosis may generate pulmonary embolism [6].

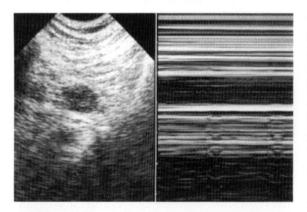

Fig. 12.15. Occlusive thrombosis of the subclavian vein, short axis. The vein is incompressible. The *right figure*, in time-motion, depicts a very sensitive sign of occlusive thrombosis: complete absence of respiratory dynamics of the vein

Ultrasound and Central Venous Catheterization

Ultrasound offers considerable help in central venous catheterization. Not only does it allow approaching the zero fault level, but it also has many effects: a considerable gain in time, more comfort for the patient, and substantial savings. Two methods are available. Ultrasound before needle insertion, which allows one to choose the most adequate of six possible sites of insertion: ultrasound-enlightened catheterization. Ultrasound during needle insertion is referred to as ultrasound-assisted catheterization.

Contribution of Ultrasound Before Internal Jugular or Subclavian Catheterization: Ultrasound-Enlightened Catheterization

The static approach alone of the vessel one intends to puncture is already rich in information. It has been proven that large veins are easier to catheterize than small ones [8]. Obviously, any catheterization should be preceded by an ultrasound verification of the vein, and every ICU should have a small simple device only for this application.

The best site can be chosen. As previously mentioned, observation shows that asymmetry is the rule at the internal jugular level. It is generally frank. A large venous lumen is sometimes associated with a contralateral very small, possibly hypoplastic one. These data have been investigated [9]. Asymmetry, defined as a cross-sectional area greater than twice that of the contralateral vein, was present in 62% of cases, to the benefit of the right side in only 68% of cases. This study also highlighted that 23% of the internal jugular veins had, at admission in the ICU, a cross-sectional area less than 0.4 cm^2 (supine patient). Systematic use of the right side thus means that a small vein will be encountered in a quarter of cases. Such a small area, which was only slightly increased by the Trendelenburg maneuver, indicated foreseeable difficulties in blind emergency catheterization.

Other disorders can explain a priori difficulties in catheterization:

- Thrombosed vein.
- Aberrant location of the vein related to the artery, which affects 8.5% of cases [10].
- Inspiratory collapse of the vein. Ultrasound provides a clear image of this situation. In this setting, no experimental studies are required to predict that there is a major risk of gas embolism here. Note that the increase in inspiratory caliper (i.e., in the sedated patient) is always correlated with a centrifuge flow of blood during disconnection of the devices. Let us specify that this route, which had the reputation of having constant dimensions even in hypovolemic patients, can be discovered to be completely collapsed when studied by ultrasound.
- Permanent complete collapse. Is it possible to even visualize such veins? Experience will often make it recognizable by very subtle handling of the probe, which should not flatten the vein. The vein is sometimes enlarged, from 0 to 1 or 2 mm at inspiration (in mechanical ventilation). The traditional Trendelenburg maneuver will not be always effective. At the internal jugular level, it is possible to compress the neck using one's free hand, just over the clavicle: a small jugular vein can then appear, but this small caliper may discourage one from the catheterization.

Central Internal Jugular or Subclavian Ultrasound-Assisted Catheterization

Blind insertion of an internal jugular or subclavian catheter failed in 10%–19% of cases, and complications occurred in 5%–11% of cases, depending on whether the operator was experienced [11]. Other studies have demonstrated that the failure rate increases with the gravity of the emergency, up to 38% in case of cardiac arrest [12]. Loss of time and complications can severely penalize the patient. Note that the physician, although next to the patient, cannot help in case of instability: the patient remains inaccessible during the entire procedure.

Permanent ultrasound guidance is mandatory when a needle is inserted in a central vein. It was described long ago [13], with many studies following that have demonstrated the advantages of ultrasound [14]. This method is of little interest to physicians who have never encountered technical difficulties. In our experience, the ultrasound-guided procedure's single drawback is its simplicity. Regardless of how clever we were before adopting this method, we have found that the ability to find any vein in a few seconds was sufficient reason for developing this technique.

Obviously, before learning ultrasound-guided catheterization, the physician should have a working knowledge of blind techniques for three reasons. First, the ultrasound unit can break down. Second, the ultrasound-assisted procedure, although very efficient, does not improve one's techniques in blind procedures, since the landmarks are completely different in both approaches. Third, one must have experienced stressful situations using the blind approach to fully appreciate the comfort that ultrasound guidance provides.

Basic details of interventional procedures can be found in Chap. 26. The probe is applied just proximal to the site of needle insertion. Asepsis must be absolute. A simple sterile glove surrounding the probe is an unacceptable solution. A 45° angle is made between probe and needle. The vein should be visualized in its long axis, and needle insertion is monitored using a longitudinal scan, aiming at the probe landmark (Fig. 12.16). Using this approach, the needle and the target are visualized over the entire length (Fig. 12.17). The artery will not appear in the screen.

Available techniques in the literature describe a system of servo control to the probe and use a transversal approach. The artery is visible beside the vein, but the progression of the needle is blind. The needle can pierce a superficial structure with impunity. Moreover, the servo control is restrictive rather than liberating in our experience. Last, the usual dedicated devices are limited to this use only, and the quality is often suboptimal. In practice, we avoid this technique.

We previously used a simple and quick method at the internal jugular vein: make a skin landmark at the area of the vein, switch off the ultrasound unit, insert the needle. However, this method was valid only if the caliper of the vein was large enough. In fact, if the internal jugular vein is large, it will be easily catheterized with blind methods [8]. To sum up, if ultrasound identifies a large vein, it will have the advantage of predicting easy catheterization using the usual blind techniques.

Note that identifying ultrasound-assisted landmarks followed by blind cannulation has been used by other teams at the subclavian level [15]. This approach seems highly hazardous since a small error in angulation will definitely result in failure. In spite of this questionable methodology, it was concluded that ultrasound was of no benefit

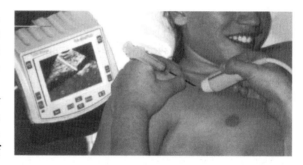

Fig. 12.16. The operator's hand holding the probe exposes the vein in its long axis and remains strictly motionless over the thorax. The operator's hand holding the needle is firmly positioned in front of the probe's landmark. The needle is then easily inserted toward the vein. For more clarity, gloves and sterile sheath are not shown in this fictitious procedure

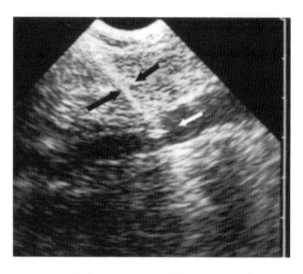

Fig. 12.17. Subclavian venous catheterization. The body of the needle is hardly visible in this scan (which does not reflect the real-time pattern) through the superficial layers (*black arrows*) and the tip of the needle has reached the venous lumen (*white arrow*)

in this setting. We think ultrasound deserves a better chance.

When should ultrasound-assisted catheterization be proposed?

- After failure of a blind attempt
- When an adequate vein is not found using ultrasound
- If costs must be controlled, since ultrasound uses 40% less material than blind techniques [16]
- In patients with official contraindications to the blind technique (see next section).
- In any situation where time must not be wasted
- More generally, if one wishes to avoid any risk or discomfort to the patient

Ultrasound-Guided Subclavian Catheterization

We prefer the subclavian route, since the risk of infection is lower. Physicians rightly fear this route, reputed to be dangerous, since immediate complications are more dramatic than in others. However, we think that the classic contraindications (impaired hemostasis, impaired contralateral lung, obesity, etc.) are no longer contraindications if ultrasound is used. Ultrasound thus benefits from all the advantages of the subclavian route with no drawbacks. In addition, thrombotic risks seem to be low, and the patient's comfort is enhanced.

In a personal study of 50 procedures carried out in ventilated patients, the success rate was 100% [17]. In 72% of cases, frank flux was obtained in the syringe in less than 20 s, in 16% of cases in less than 1 min. In 12% of cases that were considered laborious, success was nonetheless obtained in less than 5 min. In other words, ultrasound has accustomed us to immediate successes, and 5 min is considered a rather long and laborious procedure. It is crucial to specify that the patients in this study were consecutive. Among them, 13 patients were plethoric (with a distance from the skin to the subclavian vein ≥ 30 mm). They were all successfully catheterized, with an immediate procedure in 11 of them.

When the procedure is over, absence of pneumothorax (if needed) is checked using ultrasound (see Chap. 16). Checking that the catheter is not ectopic is similar. An ectopic position can also be immediately recognized during catheterization, since a metallic guidewire is perfectly visible (Fig. 12.18). It is thus wise to set the sterile sheet in order to have access to both the subclavian and the

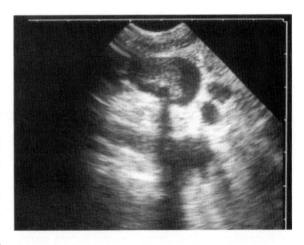

Fig. 12.18. This figure is included to show the characteristic pattern of a metallic guidewire or catheter (same pattern) in the venous lumen. This type of material generates a continuous hyperechoic mark with a frank posterior shadow. In addition, note the substantial venous thrombosis surrounding this internal jugular catheter

jugular areas. If the point of insertion of the needle is chosen rather distal to the medial line, the risk of ectopic positioning in the jugular vein decreases, as does, theoretically, any risk of subclavian pinch-off syndrome.

Ultrasound guidance at the subclavian level is also mentioned by other teams [18], but studies conducted in the intensive care setting are rare.

Real-time analysis is rich in information. One can see the needle arriving in contact with the proximal venous wall, pushing this wall, then penetrating the vein. Sometimes the proximal and distal walls are pressed against one another, and the needle pierces the vein. We have observed more dramatic phenomena: repeated puncture of a subclavian vein with a large caliper can cause a subsequent decrease in lumen size, and therefore be impossible to recognize, a chain of events that occur as if there were complete spasm. Obviously, this can only create a vicious circle that reduces the chances of success.

The needle is not always visualized during its penetration. This problem will be evoked in Chap. 26.

Ultrasound-Assisted Internal Jugular Catheterization

One can of course use the previous technique at this level, the basic rules remain unchanged [19].

Emergency Insertion of a Short Central Venous Catheter

An additional weapon can be used in the extreme emergency. Under sonographic guidance, we can insert a 60-mm, 16-gauge catheter in a central vein. In our areas, such material is, alas, difficult to find, and in practice, we make have them custom-made. Using this certainly temporary and hardly academic, but potentially lifesaving method, the problem of central venous access can be resolved in a few instants, avoiding transosseous access or others.

Can Ultrasound Replace Radiograph Monitoring After Insertion of a Central Catheter?

What do we ask of the traditional bedside radiograph? First and foremost, pneumothorax information. Ultrasound will check for absence of pneumothorax in a few seconds and with more accuracy than a bedside radiograph (see Chap. 16). Second, is the catheter in an ectopic position? Where has the catheter gone? In a majority of cases, it enters the jugular internal vein (after a subclavian insertion); ultrasound can detect this during the procedure. If it enters the cardiac cavities and the right auricle is easily visible, the end of the catheter is also visible. If not, measuring the length of the catheter to be inserted beforehand provides clinical landmarks; combined with common sense, this complication is nearly impossible.

The other causes of malpositioning are very rare. Poor outflow is a valuable clinical sign of insertion in a small-caliper vessel (a condition hard to imagine if the catheter has been inserted with ultrasound guidance). If the monitoring radiograph is requested the next day, or during a new situation, using ultrasound reduces cumulative irradiation and costs.

In practice, we no longer request follow-up radiographs and have not done so for many years [20].

Vena Cava Superior and Left Brachiocephalic Vein

The vena cava superior is looked for at the supraclavicular fossa, with the probe applied toward the neck. Generally, analysis is disappointing, because the vein is surrounded by hindering structures (lung). However, some patients have good anatomy. The Pirogoff confluent, the vena brachiocephal-

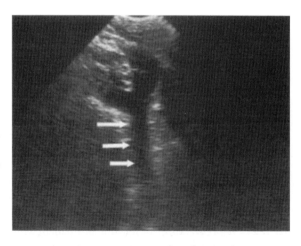

Fig. 12.19. Vena cava superior (*arrows*) whose first 3 cm are visible in this view. Depending on the quality of exposure, one can recognize the aorta inside the vein, the right pulmonary artery posterior to the vein, and sometimes the right auricle

ica, can then be recognized, and, more central, the supra-aortic trunks, the right pulmonary artery (passing posterior to the vein) and last the right auricle (Fig. 12.19).

Direct signs of venous thrombosis will be hard to detect since this area is not very accessible and cannot be compressed. Doppler can be helpful. However, several indirect signs are available to indicate good patency or an obstacle: permanent dilatation, without inspiratory collapse (in a spontaneously breathing patient) of the upper veins [21, 22]. Logically, inspiratory collapse of the subclavian vein indicates absence of an obstacle. The sniff test consists in sudden inspiration by the nose [21], which should normally yield jugular and subclavian collapse. This test is hard to carry out in the critically ill patient, since his cooperation is needed. In addition, as with any sudden maneuver, one can ask if this test is insignificant if there is, for instance, floating venous thrombosis that had been stable until then.

When the suspicion of thrombosis is high, a transesophageal examination can clearly analyze the vena cava superior.

Atelectasis is not a rare situation in the ICU. It can make the mediastinum accessible to ultrasound (see Fig. 17.11, p 121). A floating thrombosis in the vena cava superior was thus diagnosed using the right parasternal route (Fig. 12.20) in a patient with recent complete right atelectasis. The patient was promptly positioned in the right lateral decubitus, with the hope that a detached thrombus would choose the right lung.

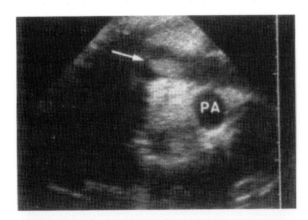

Fig. 12.20. Vena cava, superior location in a patient with right lung atelectasis. Right parasternal route. A thrombus is visible (*arrow*) within the venous lumen, and is highly mobile in real time. *PA*, right branch of the pulmonary artery

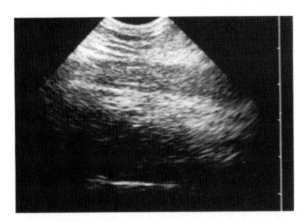

Fig. 12.21. Complete venous thrombosis at the humeral level in an ICU patient with unexplained fever. This pattern is clear when sought. Longitudinal scan at the arm

The left brachiocephalic vein is sometimes visible anterior to the aortic cross using a suprasternal route. This segment is not easy to compress. If a local thrombosis is suspected (in the case of a large left arm, for example), only static analysis will be contributive: direct detection of a thrombosis, absence of spontaneous collapse, or absence of the free valvule sign.

Upper Extremity Veins

Humeral vein thrombosis can be a source of fever after peripheral catheterization (Fig. 12.21).

General Limitations of Ultrasound

As regards the internal jugular and subclavian veins, the examination is hindered by parietal emphysema, local dressings, a tracheostomy, and cervical collars. Massive hypovolemia makes the veins hard to detect.

References

1. Wing V, Scheible W (1983) Sonography of jugular vein thrombosis. Am J Roentgenol 140:333–336
2. Dauzat M (1991) Ultrasonographie vasculaire diagnostique. Vigot, Paris
3. Chastre J, Cornud F, Bouchama A, Viau F, Benacerraf R, Gibert C (1982) Thrombosis as a complication of pulmonary-artery catheterization via the internal jugular vein. New Engl J Med 306:278–280
4. Yagi K, Kawakami M, Sugimoto T (1988) A clinical study of thrombus formation associated with central venous catheterization. Nippon Geka Gakkai Zasshi 89:1943–1949
5. Horattas MC, Wright DJ, Fenton AH, Evans DM, Oddi MA, Kamienski RW, Shields EF (1988) Changing concepts of deep venous thrombosis of the upper extremity: report of a series and review of the literature. Surgery 104:561–567
6. Monreal M, Lafoz E, Ruiz J, Valls R, Alastrue A (1991) Upper-extremity deep venous thrombosis and pulmonary embolism: a prospective study. Chest 99: 280–283
7. Ricome JL, Thomas H, Bertrand D, Bouvier AM, Kalck F (1990) Echographie avec ponction pour le diagnostic des thromboses jugulaires sur cathéter. Réan Soins Intens Med Urg 6:532
8. Lichtenstein D (1994) Relevance of ultrasound in predicting the ease of central venous line insertions. Eur J Emerg 7:46
9. Lichtenstein D, Saïfi R, Augarde R, Prin S, Schmitt JM, Page B, Pipien I, Jardin F (2001) The internal jugular veins are asymmetric. Usefulness of ultrasound before catheterization. Intensive Care Med 27:301–305
10. Denys BG, Uretsky BF (1991) Anatomical variations of internal jugular vein location: impact on central venous access. Crit Care Med 19:1516–1519
11. Sznajder JI, Zveibil FR, Bitterman H, Weiner P, Bursztein S (1986) Central vein catheterization, failure and complication rates by 3 percutaneous approaches. Arch Intern Med 146:259–261
12. Skolnick ML (1994) The role of sonography in the placement and management of jugular and subclavian central venous catheters. Am J Roentgenol 163:291–295
13. Denys BG, Uretsky BF, Reddy PS, Ruffner RJ, Shandu JS, Breishlatt WM (1991) An ultrasound method for safe and rapid central venous access. N Engl J Med 21:566

14. Randolph AG, Cook DJ, Gonzales CA, Pribble CG (1996) Ultrasound guidance for placement of central venous catheters: a meta-analysis of the literature. Crit Care Med 24:2053-2058
15. Mansfield PF, Hohn DC, Fornage BD, Gregurich MA, Ota DM (1994) Complications and failures of subclavian vein catheterization. N Engl J Med 331: 1735-1738
16. Thompson DR, Gualtieri E, Deppe S, Sipperly ME (1994) Greater success in subclavian vein cannulation using ultrasound for inexperienced operators. Crit Care Med 22:A189
17. Lichtenstein D, Saïfi R, Mezière G, Pipien I (2000) Cathétérisme écho-guidé de la veine sous-clavière en réanimation. Réan Urg [Suppl 9] 2:184
18. Nolsoe C, Nielsen L, Karstrup S, Lauritsen K (1989) Ultrasonically guided subclavian vein catheterization. Acta Radiol 30:108-109
19. Slama M, Novara A, Safavian A, Ossart M, Safar M & Fagon JY (1997) Improvement of internal jugular vein cannulation using an ultrasound-guided technique. Intensive Care Med 23:916-919
20. Maury E, Guglielminotti J, Alzieu M, Guidet B & Offenstadt G (2001) Ultrasonic examination: an alternative to chest radiography after central venous catheter insertion? Am J Respir Crit Care Med 164:403-405
21. Gooding GAW, Hightower DR, Moore EH, Dillon WP, Lipton MJ (1986) Obstruction of the superior vena cava or subclavian veins: sonographic diagnosis. Radiology 159:663-665
22. Grenier P (1988) Imagerie thoracique de l'adulte. Flammarion, Paris

CHAPTER 13

Inferior Vena Cava

Draining half of the systemic blood toward the heart and the necessary crossroads of lower extremity thromboses, the inferior vena cava (IVC) has a clear strategic situation. Ultrasound occupies a major place in the search for thromboses, but also in assessing the IVC dimensions, a possible marker of the circulating blood volume, as well as other more marginal applications.

The iliac veins are discussed in Chap. 14.

The Normal Inferior Vena Cava

The inferior vena cava can be separated by the renal veins into supra- and infrarenal portions.

The infrarenal portion analysis is conditioned by gas, frequent in this area. However, the free hand of the operator (and not the probe itself) can drive most gas away by applying gentle pressure. The suprarenal portion is often visible using the liver acoustic window. It makes its way vertically, at the right of the aorta, receives the hepatic veins and opens into the right auricle (see Fig. 4.2, p. 19). A spontaneous echoic flow can sometimes be observed. This flow can hesitate, or even be inverted at inspiration (in mechanically ventilated patients), an obvious sign of tricuspid regurgitation. This echoic flow is possibly explained by agglomerated blood cells [1] and can be massive (Fig. 13.1). Fine analysis of the content of the inferior vena cava is generally possible. Extrinsic obstacles, catheters or caval filters can be observed (Fig. 13.2).

The venous caliper is modified by respiratory and cardiac rhythms. There is usually inspiratory collapse in the spontaneously breathing subject. These variations in caliper are a sign of venous patency. A compression maneuver is perfectly possible, but the pressure should be brought by the operator's free hand with spread fingers, with the probe applied between two fingers. A compression by the probe alone would possibly damage the probe, and it can be harmful for the patient. This

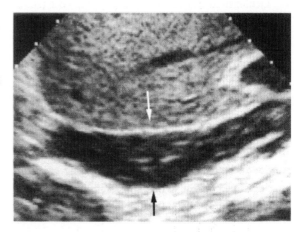

Fig. 13.1. Inferior vena cava, longitudinal scan. In this vein, an echoic flow with visible particles goes toward the right cavities. In addition, there is a bulge in the upper portion of the vein (*arrows*), a frequent variant of the normal (saber profile). Note that a measurement of the vein caliper at this level would yield misleading information in predicting central venous pressure

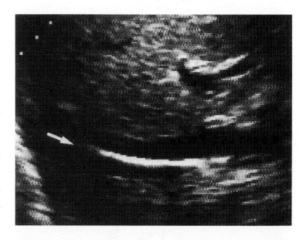

Fig. 13.2. Catheter (*arrow*) within the inferior vena cava lumen

maneuver pays off for subjects with favorable morphotype: the inferior vena cava can be easily collapsed. Note that such a maneuver does not affect the instantaneous blood pressure. The infrarenal segment can also be collapsed this way. If gentle pressure does not succeed, it seems wise not to insist.

Thromboembolic Disorders

The technique is the same as for the upper or lower extremity veins. The only difference is that the static approach should be called a pseudo-static approach, so to speak, as the frequent necessity to drive digestive gas off can alter some parameters. Thrombosis will give signs:

- Static in the static approach:
  - Endoluminal echoic irregular pattern (Fig. 13.3).
- Dynamic in the static approach:
  - Absence of spontaneous inspiratory changes (see »Normal and Pathological Patterns« below).
- In the dynamic approach:
  - Noncompressible vein. This maneuver is redundant and should not be performed if previous approaches have identified a thrombosis.

Caval Filter and Ultrasound

When local conditions are good, the correct position of a caval filter and its relations with the renal veins can be accurately assessed (Fig. 13.4).

If transportation of a critically ill patient or irradiation in a pregnant woman must be avoided, it could be advantageous to insert caval filters at the bedside, using ultrasound guidance. Once the floating infrarenal thrombus is identified, and once the indication is adequate (this would warrant an entire chapter), one operator inserts the filter while another locates the main landmarks using ultrasound. As for the pilot–bombardier relation in a B25, the two operators should be perfectly trained since the roles are permanently inversed.

The inferior vena cava can be round or flattened; see the next section.

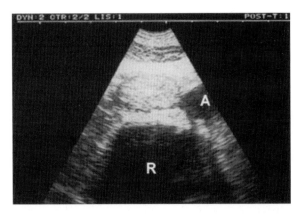

Fig. 13.3. Massive thrombosis of the infrarenal inferior vena cava. Transverse scan of the umbilical area. Anterior to the rachis (*R*) and at the right of the aorta (*A*), the venous lumen of the inferior vena cava is filled with echoic material, indicating here a recent thrombosis. Note that this recent thrombus is still soft. Hence, a compression maneuver may collapse the venous lumen, with doubtful consequences. Young patient with polytrauma

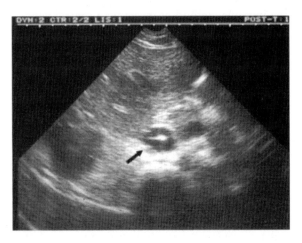

Fig. 13.4. Caval filter, perfectly identified within the lumen of the suprarenal IVC (*arrow*). Epigastric transverse scan. One can imagine the possibility of inserting this device at the bedside

Inferior Vena Cava Diameter and Central Venous Pressure

This long section gives clues for accurate measurement of the caliper of the inferior vena cava, which should take only a few seconds.

The accuracy of central venous pressure as a marker of circulating blood volume will not be discussed here. It could warrant another chapter in itself. Recently, this data has been ignored, as it appears old-fashioned to some. A discussion of modern hemodynamics can be read in Chap. 28.

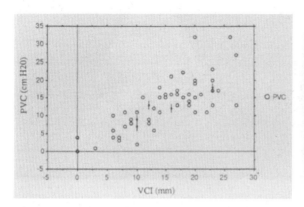 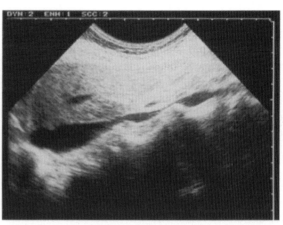

Fig. 13.5. Correlation between expiratory caliper of the inferior vena cava at the left renal vein (*VCI*) and central venous pressure (*PVC*) in 59 ventilated patients

Fig. 13.6. Irregular pattern, mostly collapsed, of the inferior vena cava. Hypovolemic patient. Note the bulge (saber profile) at the left of the image

Our aim is to provide simple noninvasive data to the intensivist who may find it useful [2]. Ultrasound measurement of the IVC caliper lies between the invasive method of inserting a central venous pressure system and the more invasive transesophageal approach.

Circulating blood volume is mainly located (65%) in the venous system. We therefore imagine that a variation in this volume will affect this sector, the IVC being an ultrasound-accessible portion. A flattened pattern in the obviously hypovolemic patients having been regularly observed, we investigated this parameter in 54 ventilated patients (Fig. 13.5). A caliper less than 10 mm was correlated with a central venous pressure under 10 cm H_2O with an 84% sensitivity, a 95% specificity, an 89% positive predictive value and a 92% negative predictive value [3]. Figure 13.5 shows that the relation is better for the small caliper values. Some studies have been conducted in this field [4–7], but most came from cardiologic, noncritical, spontaneously breathing, laterally positioned patients, with measurements made at the hepatic vein level, making any comparison difficult. Only one study dealt with ventilated patients and indicated that a caliper of $\leq$ 12 mm always predicted a central venous pressure $\leq$ 10 mmHg [7].

Measurement Technique

Simple requirements are necessary for a both accurate and reliable information.

1. The patient remains supine. Lateral decubitus would squash the IVC by the liver.
2. The IVC should be sought in a longitudinal axis first. A probably frequent mistake is the confusion between the IVC and a hepatic vein (see Fig. 4.3, p 20). Several profiles exist:
   – A regular profile.
   – A saber profile (Fig. 13.1). This frequent finding, with a bulge when the IVC receives the hepatic veins, should be recognized and the operator should remain far from this area, whose measurement would give erroneous information. In addition, the venous tissue progressively becomes cardiac tissue in this area.
   – An irregular, moniliform profile (Fig. 13.6).
3. The probe is then applied in a transverse axis. A measurement in a longitudinal axis would expose to overestimation of the caliper, when the vein is not perfectly located in a frontal axis.
4. The left renal vein should be looked for (Fig. 13.7). This landmark has two advantages: it is a reliable place, and we are definitely far from the hepatic bulge.
5. Measurement should be from face to face, not from border to border.
6. An end-expiratory measurement is needed (see »Normal and Pathological Patterns« below).
7. The increase in caliper with heart beats was not taken into account in our practice.

In addition, we did not index IVC caliper with body surface for two reasons. Risk is involved in determining these data in a critically ill, unstable patient, since it is necessary to weigh the patient. Second, IVC dimensions are not correlated with the morphotype [8]. Human eye diameter varies little in relation to weight and height as well.

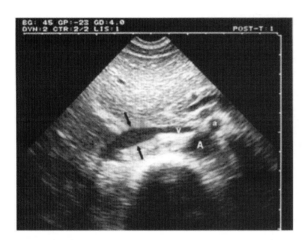

Fig. 13.7. This transverse epigastric view shows the renal veins' point of arrival. The left renal vein is particularly visible, passing between the aorta and the superior mesenteric artery (*v*), the point where we chose to measure the IVC caliper. Here, an expiratory caliper of 8 mm (*arrows*) indicates low central venous pressure

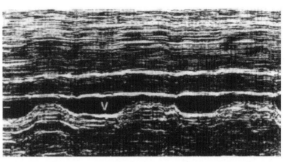

Fig. 13.8. Inspiratory collapse of the inferior vena cava. Time-motion acquisition, showing a 12-mm diastolic caliper (*V*) that collapsed to 4 mm at inspiration in a patient with major bleeding and spontaneous breathing

Normal and Pathological Patterns

In spontaneous ventilation, inspiratory caliper diminishes. This is seen in ambulatory abdominal examinations, as the patient is fasting (i.e., in moderate hypovolemia).

In mechanical ventilation, inspiratory caliper increases, for positive thoracic pressure creates an obstacle to venous return. Inspiratory collapses are found in nonsedated patients.

The expiratory caliper seems more constant. It does not vary after intubation of a patient, whereas inspiratory caliper is usually seriously disrupted.

Inspiratory collapse of a spontaneously breathing patient (Fig. 13.8) can be explained by a dyspnea with use of accessory respiratory muscles, since the inspiratory collapse of thoracic pressure creates aspiration of the systemic blood, with the Venturi effect. This situation is striking in acute asthma (where fluid therapy is not at all contraindicated). However, not all dyspneic patients, even with substantial use of accessory respiratory muscles, have inspiratory collapse.

An enlarged IVC (Fig. 7.1, p 41), with enlarged hepatic veins, is seen in right heart failure or hypervolemia, or can again be normal. Central venous pressure can be low but is rarely so.

A flattened IVC in a shocked patient (Fig. 13.7) is correlated with low central venous pressures, and indicates a hypovolemic part.

When the central venous pressure is rapidly altered, by fluid therapy, variations of PEEP or disconnecting the ventilator, IVC caliper follows (Fig. 13.9).

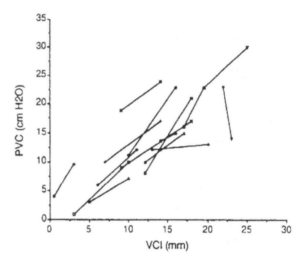

Fig. 13.9. Caliper of the inferior vena cava (*VCI*) when the central venous pressure (*PVC*) is altered

Advantages of the Ultrasonic Method

One should first note that the possible errors of this noninvasive method should be compared with the numerous errors in the measurements and interpretations of central venous pressure or wedge pressure [9]. Then the advantages can be delineated:

- These data are immediately available.
- The technique is simple (simple unit, without Doppler).
- There is no invasive procedure.
- The measurement does not affect the treatment (whereas measurement of the central venous

pressure means clamping the catheter for a short time).
- The first measurement can be used. Conversely, the first information given by central venous pressure is not very useful: the intensivist modifies therapeutic plans as its value evolves (a way to implicitly recognize the imprecision of this first value).
- A hydrostatic zero does not need to be defined, although this point can be debated. The supposed projection of the right auricle varies depending on habit. Error can be substantial in patients with widened anteroposterior thorax. Each measurement of the central venous pressure requires a number of verifications such as the height of the bed. These points cannot be checked a posteriori. We could list more of these points.
- The intensivist tries to estimate a volume (the blood volume). Central venous pressure provides a pressure (a rather indirect parameter). IVC measurement provides a distance in millimeters, which is a less indirect parameter. In discordant patients (in the right-hand column of Fig. 13.5), one can then wonder which parameter is misleading. It is in fact tempting to consider that the information given by a part of a volume is nearer the truth than the one given by a simple pressure.

In practice, this parameter will be integrated with others (heart or lung behavior; see Chaps. 17, 20, 28). We will conclude with this remark: even if a patient cannot benefit from cardiac or lung ultrasound examination, any abdominal ultrasound test performed in a critically ill patient should include the degree of IVC filling.

References

1. Dauzat M (1991) Ultrasonographie vasculaire diagnostique. Vigot, Paris
2. Magder S (1998) More respect for the CVP (Editorial). Intensive Care Med 24:651–653
3. Lichtenstein D, Jardin F (1994) Appréciation non invasive de la pression veineuse centrale par la mesure échographique du calibre de la veine cave inférieure en réanimation. Réan Urg 3:79–82
4. Mintz GS, Kotler MN, Parry WR, Iskandrian AS, Kane SA (1981) Real-time inferior vena caval ultrasonography: normal and abnormal findings and its use in assessing right-heart function. Circulation 64:1018–1025
5. Moreno F, Hagan G, Holmen J, Pryop A, Strickland R, Castle H (1984) Evaluation of size and dynamics of inferior vena cava as an index of right-sided cardiac function. Am J Cardiol 53:579–585
6. Nakao S, Come P, Mckay R, Ransil B (1987) Effects of positional changes on inferior vena caval size and dynamics and correlations with right-sided cardiac pressure. Am J Cardiol 59:125–132
7. Jue J, Chung W, Schiller N (1992) Does inferior vena cava size predict right atrial pressures in patients receiving mechanical ventilation? J Am Soc Echocardiogr 5:613–619
8. Sykes AM, McLoughlin RF, So B, Cooperberg PL, Mathieson JR, Gray RR, Brandt R (1995) Sonographic assessment of infrarenal inferior vena caval dimensions. J Ultrasound Med 14:665–668
9. Teboul JL(1991) Pression capillaire pulmonaire. In: Dhainaut JF &, Payen D Hémodynamique, concepts et pratique en réanimation. Masson, Paris, pp. 107–121

Lower Extremity Veins

The length of this chapter should in no way obscure its simplicity.

The main problem of thromboembolic disease is the risk of sudden death due to undiagnosed pulmonary embolism. The clinical data are notoriously insufficient [1, 2]. A routine test that is both accurate and applicable at the bedside is therefore of great interest. Phlebography is being increasingly replaced by ultrasound [3], which means using the Doppler mode. However, using a rigorous but simple approach without the Doppler mode, the problem can be solved in most cases. We deliberately do not discuss Doppler in this chapter.

Numerous studies have been conducted on the advantages of Doppler in thromboembolic disease. However, the studies dealing with the critically ill patient admitted in the ICU (in particular, the medical ICU) are rare. In this patient, it is illusory to consider symptomatic and asymptomatic patients.

Pulmonary embolism affects 50,000 patients yearly in France and kills 5,000 patients. Untreated pulmonary embolism has a 40% risk of mortality, angiography has a 0.04%–2% risk, and anticoagulant treatment approximately 2%.

Faced with the specter of pulmonary embolism and its various presentations, and desirous of never being surprised by this disorder reputed to be so pernicious, we have decided to no longer ask the question of whether an ultrasound examination of the venous system should be ordered. This examination is fully part of our routine examination of every admitted patient and is repeatedly performed. In some instances, the quasi-fortuitous discovery of venous thrombosis immediately clarifies situations that before were murky.

Examination Technique: Recognizing the Vein

As opposed to the ambulatory patient, the critically ill patient cannot be examined sitting, with the legs hanging down, or in ventral decubitus. Therefore, only the anterior approach in a supine patient can be routinely used. We always do transverse scans of the veins. The examination is always bilateral and comparative.

As usual, we use the same device and the same 5-MHz microconvex probe described in previous chapters. As for any vein, the probe will be held like a pen in order to control the pressure over the skin.

Quick and easy exploration is possible from the groin to the knee. The iliac and calf portions will be studied separately. The inferior vena cava was discussed in Chap. 13. A vein is recognized from:

- Anatomical topography
- The constant presence of the satellite artery, which is round, sometimes pulsatile, sometimes calcified
- Its tubular structure
- The complete flattening of the vein when pressure is applied with the probe (see next section), a major point
- Visualization of valvules at times (Fig. 14.1)

Echogenicity, on the other hand, is not informative, especially for small veins, which can be either

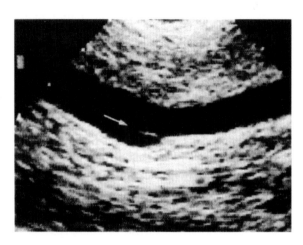

Fig. 14.1. Common femoral vein, longitudinal scan, with no compression. The *arrow* designates a valvule

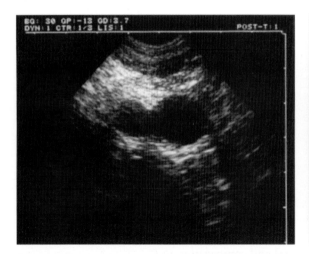

Fig. 14.2. Transverse scan of the common femoral vessels at the groin (without compression). The artery is at the *left* of the image, the vein at the *right*. The absence of apparent separation between the two vessels is due to a common tangency artifact, hence this peanut pattern

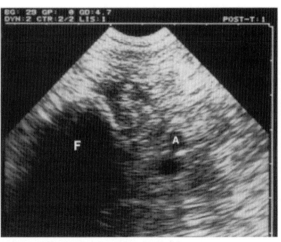

Fig. 14.3. Transverse scan of the superficial femoral vessels at the low area of the femur. The femur is an immediately recognized, large arc-like structure with frank acoustic shadow (*F*). An adequate compression maneuver identifies the vein, here under the artery (*A*)

hypoechoic or echoic if surrounded by echoic structures.

Once such a structure has been identified as a vein, it is scanned step by step to the extremity. The study can begin at the groin, since the femoral pulse makes a practical landmark. From top to bottom, the following portions will be identified:

- The common femoral vein outside the femoral pulse. At this level, vein and artery seem to be one, with a peanut pattern, since the interface, which separates vein from artery, is usually not visible (Fig. 14.2).
- The superficial femoral vein follows, vertical up to the knee, inside the femur, rarely split (Fig. 14.3).
- The deep femoral vein leaves the main femoral axis toward the femur, where it quickly disappears.
- The popliteal vein follows the superficial femoral after the Hunter canal (Fig. 14.4). The shorter the probe, the easier the popliteal vein can be explored in a supine patient.

Calf vein analysis raises problems that will be discussed in »The Calf Problem.« Other veins such as the gastrocnemial veins are not discussed.

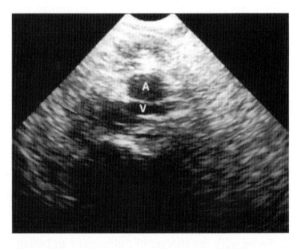

Fig. 14.4. Posterior (transversal) approach of the popliteal fossa, showing the vein (*V*), generally single, and the artery (*A*)

Examination Technique: the Compression Maneuver

A three-step approach should again be used. The two first steps are less important than at the upper extremity level.

1. Static aspect in static analysis. From the groin to the feet, the static pattern is rarely informative since the echogenicity is naturally increased.
2. Dynamic patterns in static analysis. Here again, spontaneous anomalies should be searched for, but will rarely be encountered.

3. Dynamic pattern in dynamic analysis: the compression maneuver. This is the basic maneuver, which can be used at almost every level (see Fig. 12.5). A small surface probe is particularly useful here. A transverse scan is required, since minor deviation of the axis will not make the vein disappear from the screen, as would a longitudinal scan (by the out-of-plane effect). Compression should be gently applied for three basic reasons:
   - Mild pressure is more than enough to collapse a normal vein.
   - Strong pressure can collapse an artery, especially if the blood pressure is low.
   - The safety of a high-pressure maneuver is not established. A partially detached thrombus may be dislodged [4], and ultrasound must remain a safe, noninvasive procedure. We estimate that it is wise to limit pressure to 0.5–1 kg/cm^2.

The lower femoral segment (Hunter canal) is traditionally reputed to be inaccessible to compression. This is untrue. If the free hand of the operator creates a counter-support opposite the probe, this segment can be collapsed, and even more using a moderate, controlled pressure. Some training will ensure that the operator feels where to place both hands.

The behavior of a normal vein during the compression maneuver is characteristic: the upper and lower walls get closer and eventually seem to slap against each other, resulting in a complete collapse of the venous lumen. The operator should be accustomed to feeling the necessary pressure to obtain this result. Since we did not use Doppler in our studies, this maneuver can be called the do-it-without-Doppler. For many years, using only two-dimensional ultrasound has allowed us to confirm or invalidate diagnoses of venous thrombosis. This opinion is shared by others [5, 6].

Since the first and second steps can generally be bypassed, only the compression technique is often directly performed. In this way, routine exploration of a normal femoral venous axis should not exceed 15 s. The practice of working on contiguous sections every millimeter is laudable but not very profitable, since thrombosis usually involves several centimeters of venous segment [7]. Our observations clearly confirm this notion. However, venous thrombosis is visible until the critical moment when it is no longer visible. In some cases of high suspicion of pulmonary embolism, the discovery of a centimeter-long area of femoral vein thrombosis can definitely confirm the diagnosis. We call this the miasma sign. The miasma sign is usually found at more distal areas such as the calf.

The Signs of Femoropopliteal Thrombosis

Static signs are rarely suggestive at this level, since the echogenicity of femoropopliteal veins is usually subject to the parasites of the surrounding tissues. The venous caliper can be enlarged, a possibly informative sign [8, 9]. In some instances, an echoic heterogeneous pattern is recognized within the enlarged lumen, thus making the diagnosis obvious, and the compression maneuver useless.

Dynamic signs in static analysis are not always clear, since a floating thrombosis rarely includes these narrow segments – it should be searched for at the iliac end of the thrombosis.

Conversely, the diagnosis is usually made during the dynamic maneuver. The best sign of venous thrombosis is the absence of collapse with mild probe pressure. This sign has been constant in our observations (Fig. 14.5).

The literature relates sensitivity and specificity near 100% for the diagnosis of venous thrombosis limited to this area [10]. However, studies do not specify the necessary degree of the pressure, and

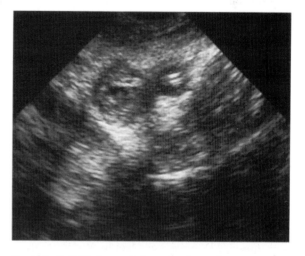

Fig. 14.5. Left iliofemoral thrombosis. In this transverse scan, registered slightly over the groin, the distal portion of the iliac vein is enlarged by an echoic heterogeneous completely occlusive material. This sole pattern renders the compression technique redundant. Hence, the risk-benefit ratio of this maneuver is inverted. Note at the *right of the image* an arterial catheter (two parallel hyperechoic lines)

they include Doppler findings. Doppler was of no use in our experience to confirm or invalidate venous thrombosis.

Other signs seem secondary. A recent thrombosis should be hypoechoic, an old one echoic. This distinction lacks reliability for some [9], and we have joined this opinion. A recent thrombosis can be deformed by the probe pressure [3]. A thrombosed vein will not be modified by a Valsalva maneuver.

Difficulties can arise from poorly echoic veins. Certain maneuvers can help in difficult cases:

- Comparing a suspect area with the contralateral area. Some patients may need more pressure than usual, in case of extreme plethora, for instance. In rare cases, however, bilateral, symmetric thrombosis can occur, and this situation can confuse the young operator.
- Filling the venous lumen when the vein seems empty:
  - Using fluid therapy.
  - Lowering the feet using the balance pedal of the bed.
  - Manually compressing the common femoral vein at the groin (after checking its patency): for the time being blood engorges the venous sector in the lower extremity.

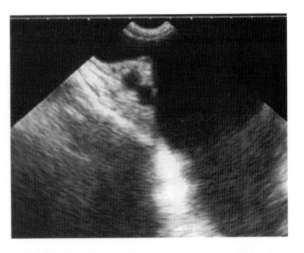

Fig. 14.6. Through a peritoneal effusion, the right iliac vessels are clearly outlined

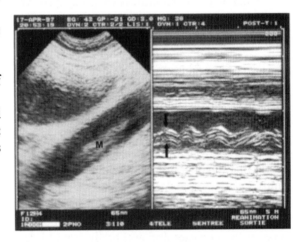

Fig. 14.7. Floating iliac thrombosis (*M*). The floating character is perfectly objectified using the time-motion mode, at the right (*arrow*). Compression of such a structure may not be sufficiently innocuous

The Iliac Level

Iliac veins are classically difficult to access with ultrasound because of the abundance of local gas. In addition, iliac segments can be incompressible even in the absence of gas. This feature is, to our knowledge, unpredictable from one patient to another. In a majority of cases, iliac veins can be followed and compressed over a more or less long portion.

In a highly echoic patient, and if one carries out the pelvic examination with both hands (one holding the probe, one gently driving away the gas), the inferior vena cava and the two iliac veins can be analyzed (Fig. 14.6). The vein can suddenly become visible, once a gas has been driven away. The pressure exerted by the free hand should drive away gas without squashing the vein (for a pseudostatic approach, see p 82). Experience alone determines the adequate pressure.

A peritoneal effusion is a fortuitous condition that makes iliac venous exploration easier: the vascular axes are isolated from the bowel.

Once the iliac veins are exposed, two approaches are available: static and dynamic. The static approach consists in directly detecting the thrombosis, if an echoic irregular tubular mass can be described within the venous lumen (Fig. 14.5) or if floating structures are identified (Fig. 14.7). The dynamic approach, i.e., the compression maneuver, is effective in some patients and completely ineffective in others. Static visualization of an obvious, more or less floating thrombosis precludes dynamic analysis, which becomes useless and dangerous. It is in the iliac or caval segments, almost never in the femoropopliteal segments, that a floating thrombosis can be observed. With experience, it becomes clear that the diagnosis of floating thrombosis is best with two-dimensional ultrasound.

Other maneuvers are possible, Doppler excepted. A Valsalva or sniff-test maneuver, in a spontaneously breathing patient – if not too tired – will either increase or collapse the iliac lumen. With mechanical ventilation, the inspiratory caliper normally increases slightly. An echoic flow with dynamic particles can again be seen at the femoral level. All these maneuvers provide information, although approximate, but which can exclude complete obstruction of the iliac or caval axes.

In some cases, the iliac veins are impossible to analyze. When is it a problem? In a non-trauma patient, isolated iliac thrombosis without femoral extension is reputed to be extremely rare, not to say nonexistent [1, 11]. In fact, if there was local catheterization, it is not rare to find iliac thrombosis with a common femoral origin. In the restricted field of thromboses occurring in pelvic disorders, for example, in the obstetrical context, and in the traumatized patient, it appears risky to set aside analysis of iliac segments. A venography or a Doppler complement should then be required. But, before indicating venography, all the possibilities of a simple technique should be used.

The Calf Problem

This section can be omitted if it is not of clinical relevance to the reader.

The main problem at the calf is that its exploration is delicate, time-consuming and risky, with no guaranteed result. Echogenicity varies from one patient to another. The veins are small and numerous (two for one artery, i.e., six veins for each leg). Their route is sinuous. At this level, operator experience is required. However, this apparent problem can be considerably tempered.

The techniques available in the literature are not applicable in the critically ill: the posterior approach in ventral decubitus, prone position, etc. A posterior approach made in a supine patient, by raising the foot, will empty the veins. Although not described in the textbooks to our knowledge, we use a method adapted to a patient immobilized in the supine position – the usual position of the critically ill. We use the anterior approach, between the tibia and fibula. In a transverse scan, both bones are easily recognized, as well as the interosseous membrane. Just anterior to it passes the anterior tibial group. Reputed to be only slightly or not at all emboligenic, this segment is usually occulted. Posterior to the membrane, through the posterior

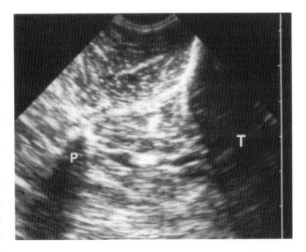

Fig. 14.8. Right calf veins, transverse scan, anterior approach. Interosseous membrane is straight between the two bones. About 2 cm posterior, through a muscle (the posterior tibial muscle), the tibial posterior and fibular veins are visible. *T*, shadow of the tibia. *P*, shadow of the fibula

tibial muscle, one can observe the fibular group outside and the tibial posterior inside (Fig. 14.8). A regular scanning by the probe from top to bottom localizes these vessels more easily. Arteries are usually not detected with our 5-MHz probe. Last, the free hand of the operator holds the calf just in front of the probe for the compression maneuvers. Sometimes, the static approach is not contributive, and only the dynamic approach makes it possible to recognize normal veins, since the collapse of small structures may be easier to recognize than the structures themselves. If needed, the head of the bed can be raised, or a tourniquet can be applied at the knee, or venous engorgement can be generated simply by compressing the common femoral vein.

Here, as elsewhere, Doppler was not used. Some authors use only the compression technique at this level, with an 85% sensitivity [12].

Basically, four scenarios are possible at the end of this exploration:

1. The three venous groups are identified. They are compressible all along the calf. One can reasonably conclude that the examination is normal.
2. A structure is identified. It is tubular, incompressible, echoic, enlarged (a very suggestive pattern if larger than 6 mm), and unilateral (Fig. 14.9). Calf thrombosis is quasi-certain. If this image is prolonged by an image clearly identified as a normal vein (the sequel sign), the

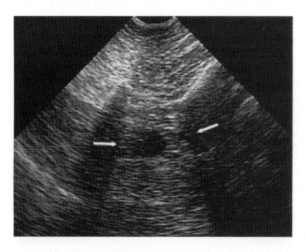

Fig. 14.9. Calf venous thrombosis. In this transverse scan, a tubular, enlarged structure is visible, at the normal place of a posterior calf vein. Above all, this structure is not compressible (*arrows*)

diagnosis of thrombosis seems certain. In a patient with acute respiratory disorder without femoral thrombosis, this finding can be decisive for the diagnosis of pulmonary embolism.
3. No tubular group is identified. No conclusion is possible. One could postulate that a thrombosed vein is enlarged, thus visible, but this remains to be confirmed.
4. At least one portion of one vein is identified and compressible. This information can be obtained in a few seconds. It already rules out massive, complete thrombosis of the leg veins. The ultrasound report will describe a calf venous system free in at least 25%, 50%, or 75% of its volume. The practical use of this approach will be discussed in the next section.

How Can the Problem of Calf Veins Be Tempered?

Calf vein detection can have an impact on the ICU's choice of equipment. Several arguments should be considered. In other words, is the calf problem a real problem?

1. In all the cases where proximal thrombosis has been detected, calf exploration is useless since the treatment is not altered. In lower extremity thrombosis, the area over the popliteal level is affected in 95% of cases [11].
2. The subpopliteal level is considered not emboligenic [13]. No fatal case of pulmonary embolism has been reported from an isolated leg venous thrombosis [14–18]. Lethal pulmonary embolism should come from the iliofemoral levels [19–21].
3. Calf thrombosis extends to the femoral veins in 20% of cases, and this extension always occurs before pulmonary embolism [18]. If this notion is taken into account, an extremely simple solution exists: when the calf level has not been well analyzed, one should monitor the distal femoral vein at the Hunter canal at regular intervals (every 24 or 48 h). A few seconds are required. If femoral thrombosis is detected by such monitoring, curative treatment can then be instigated. This procedure is conditioned by the presence or absence of a small ultrasound device at the bedside.
4. As seen above, patency of at last one part of the calf veins can be checked in a few instants. This may have immediate consequences since the risk of embolism from calf thrombosis is usually considered insignificant. If a portion of the calf veins is patent, this insignificant risk falls again and tends toward zero. Detection at any price of segmental calf thrombosis using invasive procedures or even Doppler equipment, or blind emergency treatment of isolated calf venous thrombosis will expose the patient to iatrogenic consequences. Let us then recall that pulmonary angiography is a risky examination [22], that spiral CT has a low sensitivity and that heparin therapy has an 11% risk of major bleeding and a lethal risk between 0.7% and 1.8% [23, 24]. In other words, it may not be useful to diagnose or treat in extreme emergencies calf venous thrombosis that is not massive or that is nonexistent.
5. If the physician in charge of the patient finds it essential to know the exact status of the calf veins, the gold standard remains leg venography. Experience suggests that the benefit will be small, and the drawbacks heavy (see »The Place of Venography« below).
6. Assuming that venography and blind heparin therapy both raise concerns, let us now ignore leg vein status. What then happens if there is a small thrombosis and if this thrombosis embolizes, precisely without passing by the inevitable step of simple extension? It creates a small, distal pulmonary embolism. The patient feels a low thoracic pain, but we can logically suppose that no more severe disorder will follow. This small discomfort should maybe be considered as less deleterious than the consequential drawbacks of traditional approaches. In a dyspneic patient of course, or in a patient with a poor margin of tolerance (chronic respi-

ratory insufficiency), this reasoning should possibly be nuanced. In other words, it should be assumed that there are pulmonary emboli and pulmonary emboli. Small pulmonary emboli with no residual thrombosis should possibly be considered – and managed – differently from severe pulmonary emboli as well as small pulmonary emboli with major, unstable venous thromboses.

We must remain aware that with simple logistics, the calf problem is not a true problem. The intensivist can take an interest in this segment or not, but if so, the investment will be small, since the same small equipment provides an answer to this question.

Usual Emergency Procedure

In a case of severe shock without a clear explanation, a blind fibrinolysis is sometimes planned. In such cases, the following method seems to be the quickest. Femoral axes should first be analyzed, including popliteal axes, since these segments are the most often involved, and their analysis is extremely rapid. In case of normality, internal jugular veins will be analyzed. If normal, the inferior caval and iliac veins will be included in the analysis. Lastly, subclavian veins and then calf veins will be investigated. This procedure, which may at first sight appear rather untidy, is based on logic and empiricism and will be well worth the effort from the moment a thrombosis is detected.

Limitations of Ultrasound

Iliac and calf vein analysis is uncertain.
If the compression maneuver is correct, the only false-positive cases are the rare venous tumors.
Fresh thrombosis may theoretically be compressible and yield false-negative results. The old notion of the double femoral vein (with only one thrombosed channel and the illusion of a normal single vein) has not raised any problems to date.

In a minority of cases, femoral veins cannot be recognized, in some very plethoric patients or in deep hypovolemia. Old thrombosis isoechoic to the surrounding tissues should be a limitation [16], but scanning can usually recognize a tubular structure, even isoechoic.

In the trauma patient, access can be difficult because there are numerous obstacles: orthopedic material and dressings, for example. The compression maneuver can be harmful here.

One major limitation remains that is rarely mentioned: detection of a patent vein means that there is no thrombosis, but it can also mean that there is no longer thrombosis, which is not exactly the same.

The True Place of Doppler

Information provided by the Doppler device may be largely redundant. Its high volume, high cost, high complexity, increased risk of infections (if buttons are prominent), and unproven innocuousness must be remembered. Given that the Doppler technique is highly operator-dependent, we believe that if two-dimensional and Doppler information agree, one technique is useless. If they disagree, which one should be trusted?

Doppler can be advantageous in the trauma patient, since the compression maneuver may be harmful. At the iliac level, if the clinical suspicion is real, Doppler may supersede venography. However, in case of multiple gases, it will not solve the problem.

For some, Doppler shortens the examination (an opinion we do not share) and contributes information on flow [5] or the extent of occlusion in thromboses [10]. The immediate practical use of this information seems doubtful.

The Place of Venography

Venography has a clear advantage: it provides an objective document. Certain teams still prefer venography to ultrasound, especially with young traumatized patients, where anticoagulation is never insignificant. However, venography:

- Means transportation of a critically ill patient.
- Means pelvic irradiation, iodine allergy or other accidents, and possibly pulmonary embolism.
- Transgresses (as does bedside chest X-rays) the first rule of radiology: any structure should be analyzed in two perpendicular planes. Therefore, an anterior or posterior thrombosis is easily missed in a single anteroposterior view. In addition, several areas cannot be opacified: deep femorals, twins, gastrocnemial veins, etc.

- Is operator-dependent in its interpretation. Troublesome divergences between observers are reported, from 10% to 35% [25]. Our experience confirms a high rate of errors. This is compounded if 20% [5] to 30% [26] of venographies are normal in pulmonary embolism.
- Involves a difference in cost.
- Is not a pleasant examination.

To sum up, if ultrasound has limitations, venography has other limitations, also a problem.

Interventional Ultrasound

Ultrasound can help in femoral vein catheterization in exactly the same manner as for the upper axes. A few seconds suffice to check that the vein is not thrombosed, collapsed, at an aberrant location, or when the arterial pulse is missing.

Conclusions

Our daily experience shows that compression ultrasound is a rapid, easy and reliable method. Large screening for venous thrombosis in any new or chronic patients is therefore feasible.

We usually combine lower-extremity vein analysis with examination of the internal jugular and subclavian axes. This approach provides a nearly complete overview of the deep venous axes in an acceptably short time. The alternative would be venography of the lower and upper extremities, with front and profile acquisition. Such a test is very unlikely to become routine.

References

1. Haeger K (1969) Problems of acute deep vein thrombosis: the interpretation of signs and symptoms. Angiology 20:219-223
2. Kakkar VV (1975) Deep venous thrombosis: detection and prevention. Circulation 51:8-12
3. Dauzat M (1991) Ultrasonographie vasculaire diagnostique. Vigot, Paris
4. Perlin SJ (1992) Pulmonary embolism during compression ultrasound of the lower extremity. Radiology 184:165-166
5. Cronan JJ (1993) Venous thromboembolic disease: the role of ultrasound, state of the art. Radiology 186:619-630
6. Lensing AW, Prandoni P, Brandjes D, Huisman PM, Vigo M, Tomasella G, Krekt J, Wouter Ten Cate J, Huisman MV, Büller HR (1989) Detection of deep-vein thrombosis by real-time B-mode ultrasonography. N Engl J Med 320:342-345
7. Markel A, Manzo RA, Bergelin RO, Strandness DE (1992) Pattern and distribution of thrombi in acute venous thrombosis. Arch Surg 127:305-309
8. Murphy TP, Cronan JJ (1990) Evaluation of deep venous thrombosis: a prospective evaluation with ultrasound. Radiology 177:543-548
9. Mantoni M (1989) Diagnosis of deep venous thrombosis by duplex sonography. Acta Radiol 30:575-579
10. Vogel P, Laing FC, Jeffrey Jr RB, Wing VW (1987) Deep venous thrombosis of the lower extremity: ultrasound evaluation. Radiology 163:747-751
11. Rose SC, Zwiebel JZ, Miller FJ (1994) Distribution of acute lower extremity deep venous thrombosis in symptomatic and asymptomatic patients: imaging implications. J Ultrasound Med 13:243-250
12. Yucel EK, Fisher JS, Egglin TK, Geller SC, Waltman AC (1991) Isolated calf venous thrombosis: diagnosis with compression ultrasound. Radiology 179: 443-446
13. Alpert JS, Smith R, Carlson J, Ockene IS, Dexter L, Dalen JE (1976) Mortality in patients treated for pulmonary embolism. JAMA 236:1477-1480
14. Moser KM, LeMoine JR (1981) Is embolic risk conditioned by location of deep venous thrombosis? Ann Intern Med 94:439-444
15. Appelman PT, De Jong TE, Lampmann LE (1987) Deep venous thrombosis of the leg: ultrasound findings. Radiology 163:743-746
16. Cronan JJ, Dorfman GS, Grusmark J (1988) Lower-extremity deep venous thrombosis: further experience with and refinements of ultrasound assessment. Radiology 168:101-107
17. Meibers DJ, Baldridge ED, Ruoff BA, Karkow WS, Cranley JJ (1988) The significance of calf muscle venous thrombosis. J Vasc Surg 12:143-149
18. Philbrick JT, Becker DM (1988) Calf deep venous thrombosis: a wolf in sheep's clothing? Arch Intern Med 148:2131-2138
19. Browse NL, Thomas ML (1974) Source of non-lethal pulmonary emboli. Lancet 1(7851):258-259
20. De Weese JA (1978) Ilio-femoral venous thrombectomy. In: Bergan JJ, Yao ST (eds) Venous problems. Mosby Year Book, St. Louis, p 423-433
21. Mavor GE, Galloway JMD (1969) Iliofemoral venous thrombosis: pathological considerations and surgical management. Br J Surg 56:45-59
22. Stein PD, Athanasoulis C, Alavi A, Greenspan RH, Hales CA, Saltzman HA, Vreim CE, Terrin ML, Weg JG (1992) Complications and validity of pulmonary angiography in acute pulmonary embolism. Circulation 85:462-468
23. Levine MN, Hirsh J, Landefeld S, Raskob G (1992) Hemorrhagic complications of anticoagulant therapy. Chest 102 [Suppl]:352S-363S
24. Mant M, O'Brien B, Thong KL, Hammond GW, Birtwhistle RV, Grace MG (1977) Haemorragic complications of heparin therapy. Lancet 1(8022):1133-1135

25. Couson F, Bounameaux C, Didier D, Geiser D, Meyerovitz MF, Schmitt HE, Schneider PA (1993) Influence of variability of interpretation of contrast venography for screening of postoperative deep venous thrombosis on the results of the thromboprophylactic study. Thromb Haemost 70:573–575

26. Hull RD, Hirsh J, Carter CJ, Jay RM, Dodd PE, Ockelford PA, Coates G, Gill GJ, Turpie AG, Doyle DJ, Buller HR, Raskob GE (1983) Pulmonary angiography, ventilation lung scanning and venography for clinically suspected pulmonary embolism with abnormal perfusion lung scan. Ann Intern Med 98:891–899

Pleural Effusion and Introduction to the Lung Ultrasound Technique

The pleural cavity, a basic target in the critically ill patient, is highly accessible to ultrasound. It is possible to accurately diagnose pleural effusion, to specify its nature, and to safely analyze it through direct puncture, even in a ventilated patient.

Traditionally, thoracic ultrasound is limited to the exploration of pleural effusion, with variable penetration. We will see in the following chapters that this vision can be broadened. If the indication of pleural effusion alone is considered, and even though it was described long ago [1], this application is not exploited to its fullest in all institutions. A lack of solid data may explain this paradoxical situation.

We will use this chapter to introduce the basic notions of lung ultrasound.

Basic Technique of Pleuropulmonary Ultrasonography

Lung ultrasound is a dynamic approach. It requires precise definition of the patient's situation with respect to the earth–sky axis. Fluids want to descend, gases to rise. We can thus separate lung disorders into dependent disorders, which include pleural fluid effusion and a majority of alveolar consolidations, and nondependent disorders, which include pneumothorax and generally interstitial syndrome.

The critically ill patient can be examined supine or sometimes laterally, rarely in an armchair, almost never in the prone position. Dependent lesions become nondependent if the position of the patient has changed. These features must be precisely defined during an examination, even at the price of redundancy. For instance, we describe a »posterior dependent pleural effusion in a supine patient.«

The lung surface is very large (about 1,500 cm^2). The lung is the most voluminous organ, and the question is raised of where to apply the probe. The answer could be at the same places as the stethoscope, which is perfectly realistic. In some instances, one stroke of a stethoscope answers the clinical question. For more detail, like the abdomen, the lung surface can be divided into nine well-defined areas:

1. The anterior zone (Fig. 15.1) is limited by the sternum, the clavicle, the anterior axillary line and the diaphragm. This zone can be divided into upper and lower halves, or again into four quadrants like the breast.
2. The lateral zone (Fig 15.1) extends from the anterior to the posterior axillary lines. The posterior limit, at the posterior axillary line, is thus explored with the probe at bed level in a supine patient. The bed prevents the probe from exploring more posterior areas.
3. The posterior zone (Fig. 15.2) extends from the posterior axillary line to the rachis. It can be divided into upper, middle and lower thirds, which roughly correspond to the dorsal segment of the upper lobe, the Fowler lobe and the posterobasal segment of the lower lobe.

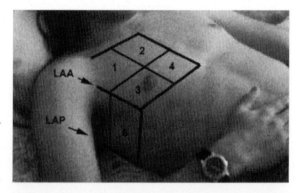

Fig. 15.1. The individualizable areas of thoracic ultrasonography. Areas 1, 2, 3, 4: superior-external quadrant, etc. of the anterior aspect. Areas 5 and 6: upper and lower areas of the lateral aspect. *LAA(P)*, axillary anterior (posterior) line

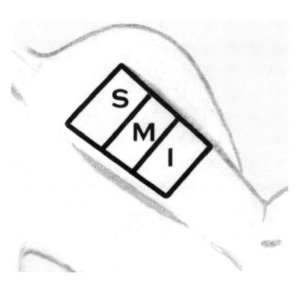

Fig. 15.2. Upper (*S*), middle (*M*) and lower (*I*) areas of the posterior pulmonary aspect. The patient can be in the ventral decubitus, but is usually in the lateral position for this analysis, and can even remain in the dorsal decubitus if the probe is short (see Fig. 15.3)

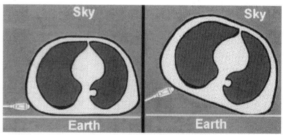

Fig. 15.3. On the left figure, the probe explores the lateral zone up to bed level. The bed prevents the probe from going further. On the right figure, the back of the patient has been slightly raised (the lateralization maneuver), and the probe then reaches precious centimeters of exploration. Minimal effusion or very posterior consolidation can be diagnosed. Note that the probe, with respect to the horizon, is pointed toward the sky

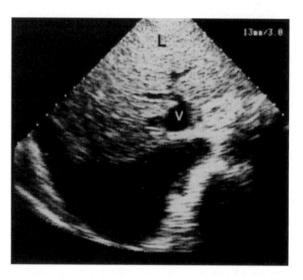

Fig. 15.4. Pleural effusion as it appears during a transabdominal approach, through the liver (*L*), in a transversal scan. This traditional approach does not provide a definite diagnosis with certain lower-lobe consolidations and also does not allow ultrasound-guided thoracentesis. Note that the effusion goes posterior to the inferior vena cava (*V*), a feature that distinguishes, if necessary, pleural from peritoneal effusion

In practice, stages of investigation can be defined:

- Stage 1. Supine analysis of the anterior wall alone defines investigation stage 1. This approach detects or rules out pneumothorax and interstitial syndrome in a few instants.
- Stage 2. Addition of the lateral zone to the anterior zone immediately detects clinically relevant pleural effusions and alveolar consolidations. We sometimes speak of pleural effusion detectable when the bed prevents further progression of the probe.
- Stage 3. To examine at least a portion of the posterior zone in a supine patient, the patient is slightly rotated, by taking the arm to the contralateral shoulder (Fig. 15.3). This slight rotation allows a short probe to be inserted as far as possible and explore a few centimeters of the posterior zone. The probe should point to the sky. This lateralization maneuver defines stage 3. The small pleural effusions and alveolar consolidations that were not detected by the previous maneuvers become accessible. Subposterior effusion implies that the patient remained supine and underwent the lateralization maneuver.
- Stage 4. This stage implies substantial analysis, including analysis of the posterior zones after positioning the patient in the lateral decubitus. An analysis of the apex will be added, by applying the probe at the supraclavicular fossa in a supine patient. Stage 4 offers more information, which makes ultrasound nearly as competitive as CT, as will be proven [2].

The intercostal spaces are always directly explored. We never use the traditional subcostal approach, which appears insufficiently informative, not to say sometimes misleading (Fig. 15.4). Our small microconvex probe is perfect for the intercostal approach.

The practice of longitudinal scans makes it possible to always keep the ribs under visual control, a

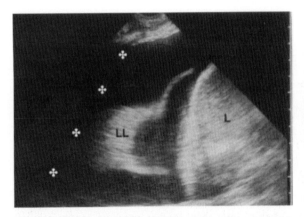

Fig. 15.5. Substantial pleural effusion by the intercostal route, longitudinal scan of the right base. Principal features are the anechoic pattern of the effusion, which just evokes the transudate. The lower lobe (*LL*) is swimming within the pleural effusion in real-time. The hemidiaphragm, located just above the liver (*L*), moves in rhythm with respiration, its course can be clearly measured. The posterior shadow of a rib (*asterisks*) hides a portion of the alveolar consolidation. Note that the pleural effusion and this posterior shadow are both anechoic. This anechoic area is real for the former and artifactual for the latter

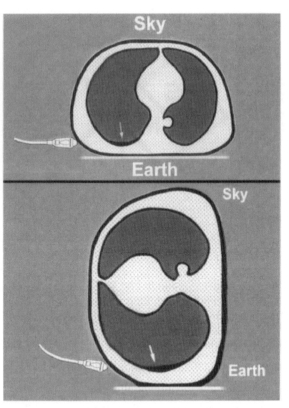

Fig. 15.6. This minimal effusion follows the laws of gravity. It is impossible to detect since the probe points downward to the center of the earth, regardless of whether the patient is studied at bed level (*top figure*) or in the lateral decubitus (*bottom figure*)

basic landmark (the bat sign; see Fig. 16.1, p 105) in order to avoid serious mistakes. The next step is to locate the thorax: one must therefore locate the diaphragm, which can be visible through a pleural effusion (Fig. 15.5) or not visible (see Fig. 4.9, p 22). The diaphragm is usually recognized: a large hyperechoic concave structure which descends – in principle – at expiration. Everything above (i.e., at the left of the image) is thoracic, everything under is abdominal. This precaution avoids confusion between pleural and peritoneal effusions, and also between alveolar consolidation and normal abdominal structures. The diaphragm, in a supine patient, is located most often at the mamillary line or a few centimeters below.

The following step, fine analysis of the pleural layers, will be detailed in Chap. 16.

Normal Aspect of the Pleura

The pleural cavity is normally virtual. Distinguishing between parietal and visceral layers is not possible using a 5-MHz probe, but this limitation is without clinical relevance. At the pleural line (which will be described in more detail in Chap. 16), the only visible elements are lung sliding and air artifacts, which belong to the group of lung signs, to be studied in Chaps. 16 and 17. Figures 16.1–16.3, pp 105–106, and 17.6–17.9, pp 120, all correspond to normally joined pleural layers.

Positive Diagnosis of Pleural Effusion

The first ultrasound description of pleural effusion seems to have been made in 1967 [1]. We should immediately point out a basic detail: pleural effusion collects in dependent areas. Any free pleural effusion will therefore be in contact with the bed in a supine patient. This zone will not be easy to approach. Rotating the patient in the lateral decubitus will not be entirely satisfactory, since the effusion will subsequently move (Fig. 15.6). The main key to detecting the effusion is to give a maximal skyward direction to the probe, which is inserted to its maximum at the (supine) patient's back, thus using the lateralization maneuver and pointing as much as possible toward the sky. Therefore, a long probe will be a major hindrance

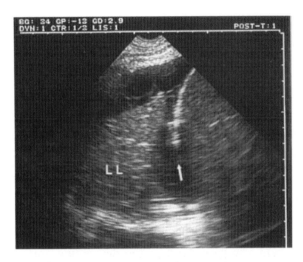

Fig. 15.7. This scan is not very different from that of Fig. 15.5. However, the effusion is less voluminous and septations are visible. The lower lobe (*LL*) is entirely consolidated. In this patient with purulent pleurisy, in real-time the hemidiaphragm was completely motionless

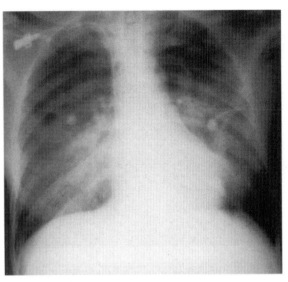

Fig. 15.9. Bedside radiography performed in a patient with acute respiratory failure. The initial diagnosis was cardiogenic pulmonary edema. Both cul-de-sacs are free, thus indicating absence of pleural effusion. However, not only pleural effusion was proven using ultrasound, but 20 cc of effusion were safely withdrawn in this mechanically ventilated patient. Immediate analysis of the fluid indicated exudate, a finding which modified the immediate management (definitive diagnosis was infectious pneumonia)

to this maneuver, and to the practice of lung ultrasonography in the critically ill.

In our experience, the diagnosis of pleural effusion depends on static and above all dynamic signs. The main static sign is the detection of a dependent collection, limited downward by the diaphragm, superficially by a regular border, the parietal pleural layer, always located at the pleural line, and deeply by another regular border, the visceral pleural layer (Fig. 15.7). The more reliable sign is in our experience dynamic: the deep border, which indicates the visceral pleural layer,

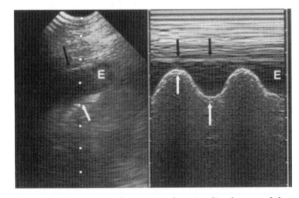

Fig. 15.8. The sinusoid sign. In a longitudinal scan of the base, this collection's thickness (*E*) varies in rhythm with the respiratory cycle. The deeper border (*black arrows*) moves toward the chest wall, thus shaping a sinusoid, whereas the superficial border (*black arrows*), which designates the pleural line, is motionless. The sinusoid sign is specific to pleural effusion

moves toward the parietal pleura at inspiration (Fig. 15.8). This sign, which could be called the sign of the respiratory interpleural variation, or the sinusoid sign, is mandatory for an accurate diagnosis of pleural effusion. Its specificity is 97% [3]. The visualization of a floating and freely rippling lung within the collection, like a jellyfish (the jellyfish sign), is a variant of this sign (Fig. 15.5). The sinusoid sign affords two advantages: first, it is specific to pleural effusion. Second, it indicates low viscosity, as we will see below. In very viscous effusion or septate effusion, the sinusoid sign is not present. Note that a complex echostructure is a criterion of fluid collection [4].

Ultrasound provides many advantages compared to the physical examination (we rarely hear a pleuritic murmur or pleural rubbing in critically ill patients), but above all compared to radiography (Fig. 15.9). Ultrasound is recognized as the choice method to detect pleural effusion in a supine patient [5]. It usually detects the effusion that is occulted in radiography [4]. Up to 500 ml can be missed with bedside radiography [6, 7]. We will see that ultrasound can diagnose and even safely tap pleural effusion that is radio-occult, even

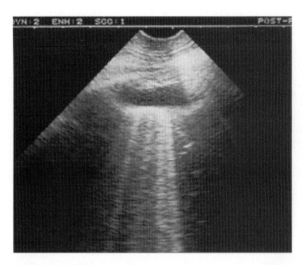

Fig. 15.10. Minimal pleural effusion, longitudinal scan of the base, patient slightly rotated to the contralateral side. In this scan, the distance between skin and parietal pleura (16 mm) can be accurately measured. The interpleural inspiratory distance is 7 mm, a finding that discourages a diagnostic tap. The air artifacts posterior to the effusion indicate an absence of alveolar consolidation at this level. If the probe placed at the anterior aspect of the chest wall (in a supine patient) showed the same pattern, this would indicate major pleural effusion

in a ventilated patient [3]. Conversely, when the radiograph is very pathological, ultrasound distinguishes the fluid and the solid components. When directly comparing ultrasound to CT, specificity of ultrasound is 94% and specificity 86% – a rate that increases up to 97% if only effusions over 10 mm thick (i.e., a very low threshold) are taken into account [2]. In brief, the majority of missed effusions are minimal effusions. Paradoxically, ultrasound can perfectly detect effusions on the millimeter scale (Fig. 15.10), provided the probe is applied at the right spot, which can be difficult with respect to the constraints of gravity (see Fig. 15.6).

Evaluating Pleural Effusion Quantity

A pleural effusion lies in the dependent part of the chest.

Minimal effusion will be detected only at the posterior aspect in a supine patient (Fig. 15.10). The more the effusion is abundant, the more anterior it will be detected (in a supine patient), at the lateral wall, then at the anterior wall (see Fig. 15.5). Detection of minimal effusion at the anterior wall (in a supine patient) assumes abundant effusion.

An aerated lung floats over the effusion, whereas a consolidated lung has the same density and swims as if in weightlessness (jellyfish sign).

With experience, and without yet being able to provide a reliable key, the rough volume of the effusion can be appreciated, if one accepts a wide margin. For instance, an effusion will contain between 30 and 60 ml, or between 1,000 and 1,500 cc. This approximation seems more precise than the words »minor«, »moderate«, etc. A possible landmark can be the location where the effusion begins to be visible.

Note that abundant effusion will allow analysis of the deeper structures such as the lung if consolidated, the mediastinum, the descending aorta, etc. One should take advantage of this effusion to quickly explore these deeper structures before evacuation, except in an emergency.

Diagnosis of the Nature of Pleural Effusion

In the ICU, the main causes of effusion are transudate, exudate, and purulent pleurisy. In a medical ICU, Mattison found a 62% prevalence of pleural effusion, 41% at admission [8]. The main causes were cardiac failure (35%), atelectasis (23%), parapneumonic effusion (11%) and empyema (1%). Analysis of the echogenicity provides a first orientation [9]: to sum up, all transudates are anechoic, anechoic effusions can be either transudates or exudates, and all echoic effusions are exudates. However, our observations show that it is more advisable to go further still. Only thoracentesis will provide an accurate diagnosis.

Transudate

A transudate yields completely anechoic effusion. This can be difficult to assess if the conditions are poor, with parasite echoes in plethoric patients, for instance. When the conditions make evaluation feasible, an anechoic effusion should not systematically be punctured in the appropriate clinical context, i.e., when there are no infectious signs, in a patient with positive hydric balance, etc.

Exudate

Exudate can be anechoic, regularly echoic, or contain various amounts of echoic particles or septations. The effusion surrounding pneumonia can have this pattern.

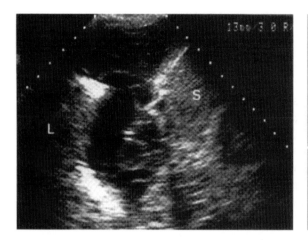

Fig. 15.11. Massive honeycomb compartmentalization in a man with pleural pneumopathy due to *Clostridium perfringens*, with septic shock. White lung on chest radiograph. *L*, lung. *S*, spleen

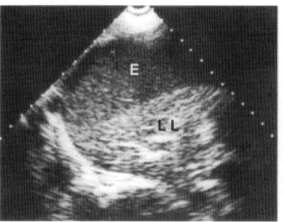

Fig. 15.12. In this exceptionally transverse view of the lateral chest wall, a complex pattern is observable. The pattern is tissular in the lower lobe (*LL*) as well as in pleural effusion (*E*). However, the *LL* area is motionless apart from the hyperechoic punctiform images that have inspiratory expansion (a sign of alveolar consolidation). The *E* area has a massive, slight movement, as would plankton in weightlessness, a sign indicating a fluid origin. The plankton sign also indicates that the effusion is an exudate and is rich in particles. Purulent pleurisy. There is no sinusoid sign, the hemidiaphragm is motionless, both findings correlated with a fall in compliance of the lung

Purulent Pleurisy

The diagnosis is usually immediately evoked since the effusion is echoic. Several cases are possible. Fine septations can be clearly observed (Fig. 15.7). These fibrin formations are nearly always missed by CT [10]. They indicate purulent pleurisy but can sometimes be seen in noninfectious effusions. The effusion can contain multiple alveoli in a honeycomb pattern (Fig. 15.11). Last, the effusion can be frankly echoic and tissue-like (Fig. 15.12). We often observe a characteristic sign that can be called the plankton sign: visualization, within an apparently tissular image, of a slow whirling movement of numerous particles, as in weightlessness. This movement is punctuated with respiratory or cardiac movements. This pattern, even discrete, indicates the fluid nature of the image. Hyperechoic elements should correspond to infectious gas.

In acute pachypleuritis due to pneumococcus, the effusion is separated from the wall by an echoic, heterogeneous thickening, tissue-like, and without sinusoid interpleural variation (Fig. 15.13).

Of course, in all these cases, the radiological pattern only shows nonspecific pleural effusion (when indeed it shows it).

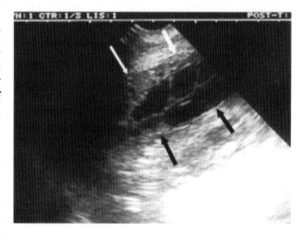

Fig. 15.13. Pachypleuritis 30 mm wide (*arrows*) in pneumopathy due to pneumococcus. Note the echoic, tissular zones, and the anechoic zones (fluid septations). Lung sliding was completely abolished

Hemothorax

Hemothorax yields echoic effusion giving the plankton sign (Fig. 15.14).

Pitfalls

An image appearing through the diaphragm during an abdominal approach is far from meaning pleural effusion. Compact alveolar consolidation can yield this pattern. The sinusoid sign can be very hard to detect by the abdominal approach.

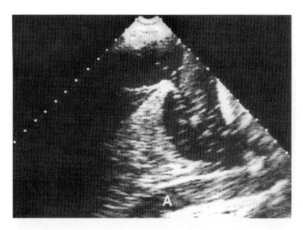

Fig. 15.14. Voluminous left hemothorax. Note in this lateral longitudinal scan showing substantial effusion that there are multiple echoes, mobile and whirling in real-time like plankton. The lower lobe is consolidated. Note through this disorder a perfectly visible descending aorta (*A*)

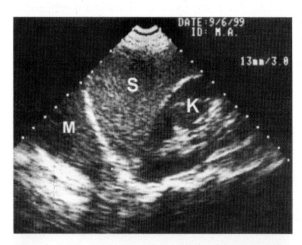

Fig. 15.15. On this longitudinal subcostal scan, the left kidney (*K*), the spleen (*S*), the hemidiaphragm, then an area (*M*) evoking pleural effusion can be observed. This is a pleural ghost generated by the spleen, which is reflected by the diaphragm, a concave reflective structure. Curiously, this mass *M* has a structure a bit too close to the spleen. The use of direct intercostal scans will make it possible to avoid this pitfall

Subphrenic organs such as the spleen can appear through the diaphragm which, like all concave structures, has reflective properties and can reverberate underlying structures at an apparently upper location. Here again, no sinusoid sign can be observed (Fig. 15.15).

An image without the sinusoid sign can be an alveolar consolidation, a very viscous or encysted effusion at the periphery of a lung which has lost its compliance.

The Ultrasound Dark Lung

In rare cases, the image is entirely hypoechoic. No difference in structure can be observed between compact alveolar consolidation and pleural effusion. Discriminant signs such as the sinusoid sign, plankton sign or dynamic air bronchogram (see Chap. 17) can be absent and prevent any conclusion. Usually, the radiological pattern is that of a white lung. This pattern is more often due to pleural effusion. In these rare cases, CT can give valuable information on whether to carry out thoracentesis.

Interventional Ultrasound

Ultrasound has the noteworthy merit of allowing puncture of an even minimal pleural effusion. Five vital organs can be recognized and avoided: heart, lung, descending aorta, liver and spleen. What is more, the rate of failure drops to zero.

Technique

An ultrasound-guided procedure can be undertaken. It is usually much simpler and effective to make an ultrasound landmark immediately before thoracentesis. The idea is to puncture where fluid is seen in a sufficient amount. The required criteria for safe thoracentesis are presence of a sinusoid sign, an inspiratory interpleural distance of at last 15 mm, visible over three intercostal spaces, and particular care taken to maintain the patient in strictly the same position for thoracentesis as for locating the ultrasound landmark [3].

The patient could be positioned in a sitting position or in lateral decubitus. In 49% of cases, it is possible to proceed in the supine position if the previous criteria are observed at the lateral chest wall [3]. The procedure here is very simple. The organs to be avoided should be located. Note than the lung may eventually appear on the screen only at the end of inspiration. In this case, another site, more dependent, should be chosen. If no safe approach is recognized, one must rotate the patient in the lateral decubitus and proceed to a posterior tap.

An ultrasound-guided tap of pachypleuritis makes it possible to aim for fluid areas. If numerous fibrin septations are observed, tap failures can be explained.

We use a 21-gauge (green) needle for diagnostic taps, and a 16-gauge (gray) needle for evacuation (see Chap. 26 for more details).

Safety of Thoracentesis

A recurrent question is the opportunity and the risk of thoracentesis in a ventilated patient. Few studies have responded to this question. In our experience, ultrasound accurately showed pleural effusion and the organs not to be punctured (see Figs. 15.5, 15.10, 15.11 and 15.14). In a study on 45 procedures in ventilated patients, the success rate was 97%, no complications occurred, in particular pneumothorax, we were able to leave the patient in the supine position in 49% of the cases, and a small-caliber needle was used each time with success: 21-gauge needles in 38 cases, and 16-gauge needles in six cases [3].

As regards evacuation thoracentesis, large tubes are usually used. These procedures are rather invasive. We always prefer to use a system we have developed with a 16-gauge, 60-mm-long catheter. This system has numerous advantages, simplicity being the first (no large wound made at the chest wall, no bursa, no risk of superinfection, minimal pain, cost savings). Ultrasound guides needle insertion, fluid withdrawal and simple catheter withdrawal at the end of the procedure with just a simple dressing applied. Using a 60-ml syringe, the fluid is withdrawn with an average flow of 1 ml/s, i.e., 20 min for a 1.2-l effusion. This corresponds to a global time saving, since no time is required for dissecting the wall, preparing the pouch with skin materials and other additional procedures. Since there is no lateral hole, the catheter should be withdrawn little by little during the procedure until it comes out of the pleural cavity. Transparent dressing is desirable, since ultrasound can better monitor the situation if a substantial amount of fluid remains.

Ultrasound gives access to approaches that would be inconceivable with only clinical landmarks. For instance, encysted pleural effusions located in full hepatic dullness have been successfully withdrawn. The liver was shifted downward.

Indications

Now that we know that thoracentesis under mechanical ventilation is a safe procedure, one can ask whether it is useful.

Diagnostic thoracentesis provides a variety of diagnoses: purulent pleurisy, hemothorax, glucothorax. Distinction between exudate and transudate has clinical consequences. The bacteriological value of a microorganism detected in a pleural effusion is definite [11]. A routine ultrasound examination at admission for all acute cases of pneumonia should theoretically allow bacterial documentation and should replace the probability antibiotic therapy. Personal observations of all patients having had thoracentesis have found an extremely high rate of positive bacteriology: up to 16%, a rate which cannot but increase if not yet treated rather than treated patients are included. Since the risk is extremely low in our experience, the high risk-benefit ratio speaks for a policy of easy puncture.

Therapeutic thoracentesis is recommended if one accepts that fluid withdrawal improves the respiratory conditions of the critically ill patient [12, 13].

Pneumothorax

Chapter 16 is devoted to pneumothorax.

References

1. Joyner CR, Herman RJ, Reid JM (1967) Reflected ultrasound in the detection and localization of pleural effusion. JAMA 200:399–402
2. Lichtenstein D, Goldstein I, Mourgeon E, Cluzel P, Grenier P & Rouby JJ (2004) Comparative diagnostic performances of auscultation, chest radiography and lung ultrasonography in acute respiratory distress syndrome. Anesthesiology 100:9–15
3. Lichtenstein D, Hulot JS, Rabiller A, Tostivint T, Mezière G (1999) Feasibility and safety of ultrasound-aided thoracentesis in mechanically ventilated patients. Intensive Care Med 25:955–958
4. Menu Y (1988) Echographie pleurale. In: Grenier P (ed) Imagerie thoracique de l'adulte. Flammarion Médecine-Science, Paris, pp 71–88
5. Doust B, Baum JK, Maklad NF, Doust VL (1975) Ultrasonic evaluation of pleural opacities. Radiology 114:135–140
6. Müller NL (1993) Imaging the pleura. State of the art. Radiology 186:297–309
7. Collins JD, Burwell D, Furmanski S, Lorber P, Steckel RJ (1972) Minimal detectable pleural effusions. Radiology 105:51–53
8. Mattison LE, Coppage L, Alderman DF, Herlong JO, Sahn SA (1997) Pleural effusions in the medical

ICU: prevalence, causes and clinical implications. Chest 111:1018–1023
9. Yang PC, Luh KT, Chang DB, Wu HD, Yu CJ, Kuo SH (1992) Value of sonography in determining the nature of pleural effusion: analysis of 320 cases. Am J Rœntgenol 159:29–33
10. McLoud TC, Flower CDR (1991) Imaging the pleura: sonography, CT and MR imaging. Am J Roentgenol 156:1145–1153
11. Kahn R J, Arich C, Baron D, Gutmann L, Hemmer M, Nitenberg G, Petitprez P (1990) Diagnostic des pneumopathies nosocomiales en réanimation. Réan Soins Intens Med Urg 2:91–99
12. Talmor M, Hydo L, Gershenwald JG, Barie PS (1998) Beneficial effects of chest tube drainage of pleural effusion in acute respiratory failure refractory to PEEP ventilation. Surgery 123:137–143
13. Depardieu F, Capellier G, Rontes O, Blasco G, Balvay P, Belle E, Barale F (1997) Conséquence du drainage des épanchements liquidiens pleuraux chez les patients de réanimation ventilés. Ann Fr Anesth Réanim 16:785

Pneumothorax and Introduction to Ultrasound Signs in the Lung

This admittedly rather long chapter will demonstrate that ultrasound can be of great assistance with an old problem: pneumothorax. A majority of pneumothoraces can be ruled out or confirmed at the bedside in just a few instants.

In addition, we will use pneumothorax as an introduction to the analysis of the normal lung, as it can be viewed as an ultrasound examination of the »non-lung.«

Introduction

Pneumothorax is a daily concern in an ICU, with a rate estimated at 6% [1], and involves a number of issues. Pneumothorax occurring under mechanical ventilation is a severe complication requiring immediate diagnosis. It is known that high-risk patients call for exceptional care, since the risk of a missed pneumothorax can be considerable [2]. On the other hand, excessive searches for pneumothorax are frequent and result in increased irradiation, delay and costs. A bedside chest radiograph does not rule out pneumothorax. Up to 30% of pneumothorax cases are occulted by the initial radiograph [3–6]. Half of these cases will become tension pneumothoraces [3]. Even a tension pneumothorax can remain unclear in a bedside radiograph [7]. In addition, in this dramatic situation, time lacks for radiological confirmation [8]. CT is the usual gold standard [9]. However, it cannot be immediately obtained without very serious drawbacks in the ICU.

Ultrasound provides an elegant answer to all of these problems.

The Normal Ultrasound Pattern of the Lung

The Pleural Line

It is traditionally considered that since the lung is an aerated organ, it cannot be investigated using ultrasound. This assertion should be nuanced. First of all, it is already possible to determine a normal pattern, made up of both static and dynamic signs. Mastering the normal picture should be acquired before any incursion into the pathological domain.

A first step will be the recognition of the ribs and their acoustic shadow in a longitudinal scan. Neglecting this step can cause serious mistakes. A hyperechoic, roughly horizontal line is located approximately 0.5 cm below the rib line: the pleural line (Fig. 16.1). The pleural line reflects the inter-

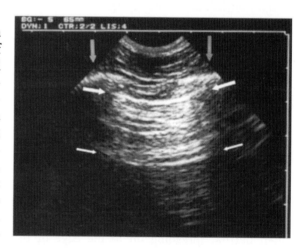

Fig. 16.1. This is the visible pattern when a probe is applied in a longitudinal axis over the thorax of a normal subject. At first sight, only artifacts are shown in this image (air artifacts surrounded by bone artifacts). The superficial layers are visible at the top of the screen. The ribs (*vertical arrows*) are recognized by their arciform shape with posterior acoustic shadow. Below the rib line (0.5 cm below), this roughly horizontal hyperechoic line (*large horizontal arrows*) is the pleural line. It indicates the lung surface. The upper rib–pleural line–lower rib profile shapes a sort of bat flying toward us, hence the bat sign, a basic landmark in lung ultrasonography. One can see a deep repetition of the pleural line (*small arrows*), the A line. This line is located at a precise place, which is the distance between the skin and the pleural line. The pleural line and the A lines are thus precisely located and should not be confused with other horizontal lines located above or below

face between the soft tissues (rich in water) of the wall and the lung tissue (rich in air). The pleural line is called the lung-wall interface. The pleural line is distinct from the aponeurotic layers and from the repeated lines in depth, since it is the only structure located 0.5 cm below the rib line (see Fig. 16.1). A bat can be imagined flying toward us, with the wings as the ribs and the back the pleural line (the bat sign).

All lung signs arise at the very level of the pleural line, which represents the parietal pleura in all cases and the visceral pleura in the cases where it is present against the parietal pleura.

Static Signs

The static signs are defined by the artifacts arising from the pleural line. They are numerous and their description would have yielded unwieldy labels. For practical purposes, they were given short names using an alphabetic classification [10].

The most clinically relevant artifacts are either roughly horizontal or roughly vertical.

The most usual artifact is a roughly horizontal, hyperechoic line, parallel to the pleural line and arising below it, at an interval that is exactly the interval between skin and pleural line. This artifact was called the ultrasound A line (see Fig. 16.1). As a rule, several A lines are visible at regular intervals. They can be called A1 lines, A2 lines, etc., according to the number of observed lines (their exact number has no clinical relevance, provided there is at least one A line).

The second by order of clinical relevance is a comet-tail artifact, roughly vertical, arising from the pleural line, well defined like a laser ray, most often narrow, spreading up to the edge of the screen without fading (i.e., 17 cm on our unit's largest scale), and synchronized with lung sliding (which will be described in »Dynamic Signs«). This precise artifact has been called the ultrasound B line (Fig. 16.2). This term may lead to confusion with the familiar Kerley B lines, but Chap. 17 shows that this analogy is not completely fortuitous. When several B lines are visible in a single scan, the pattern evokes a rocket at lift-off, and we have adopted the term »lung rockets.«

A certain vertical comet-tail artifact should in no case be confounded with a B line. It also arises from the pleural line but is ill defined, not synchronized with lung sliding, and above all, rapidly vanishes, after 1–3 cm (Fig. 16.3). This artifact has been called the Z line, the last letter of the alphabet

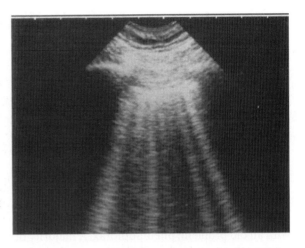

Fig. 16.2. In this scan, the superficial layers, the ribs and the pleural line described in Fig. 16.1 are present. On the other hand, artifacts arising from the pleural line here have a roughly vertical orientation and are comet tails, with well-defined, laser-like lines (seven comet tails can be counted), above all spread up to the edge of the image without fading. These are B lines, here gathered in lung rockets. These artifacts indicate that the pleural layers are correctly pressed again

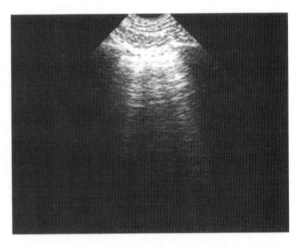

Fig. 16.3. Arising from the pleural line, three vertical, ill-defined artifacts, fading after a few centimeters can be defined. These are Z lines, a type of air artifact that should in no case be confused with B lines

symbolizing the place this artifact should take, since it has no known clinical use. One must describe another critical difference between B and Z lines. B lines erase A lines, whereas Z lines do not (see Fig. 16.2 and 16.3).

Another kind of vertical artifact should be opposed to B lines. This artifact, again a comet-tail, is well defined and spreads up to the edge of the screen without fading. However, this artifact does not arise from the pleural line but from superficial layers, and results in erasing the pleural line. The bat sign is no longer visible. This artifact has been called the E line, E for emphysema (see Fig. 16.11). We will see that parietal emphysema (or sometimes parietal shotgun pellets) can generate this artifact, which can mislead the young operator.

In some cases, no horizontal or vertical artifact is visible arising from the pleural line, and this pattern is called the O line (or the non-A non-B line). The meaning of O lines is under investigation. For the time being, they should be considered as A lines.

C lines are curvilinear, superficial images. They are described in Chap. 17.

Other types of artifacts exist (I, S, V, W and X lines), but will not all be detailed in the present edition.

Dynamic Signs

Lung sliding is the basic dynamic sign.

Description

Careful observation of the pleural line shows a twinkling at this level, in rhythm with respiration: lung sliding. In order to objectify lung sliding, we used the time-motion mode. The characteristic pattern obtained, which recalls a beach, can be called the seashore sign (Fig. 16.4). The time-motion mode provides a definite document, whereas a single frozen image cannot indicate whether lung sliding is present. Not only is a TM-mode figure easier to insert in a medical file than a video tape, but this mode helps the beginner to become aware of lung sliding. With experience, only the two-dimensional mode is sufficient.

When lung rockets are associated with lung sliding, a very frequent pattern, they behave like a pendulum that amplifies lung sliding and facilitates its perception.

With experience, 1 s suffices to recognize lung sliding, a crucial advantage.

Significance of Lung Sliding

Lung sliding shows the sliding of the visceral pleura against the parietal pleura, hence the inspiratory descent of the lung toward the abdomen. Ultrasound, a very high-precision method, is able to detect this fine movement.

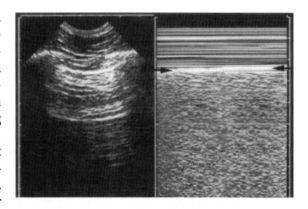

Fig. 16.4. The seashore sign. The left image is static and lung sliding cannot be identified. The right image, acquired in time-motion mode, clearly shows a double-component pattern separated by the pleural line (*arrows*). The *top* is made up of a succession of horizontal lines, recalling the sea. The *bottom*, grainy in aspect, recalls the beach, hence the seashore sign. The time-motion mode thus objectifies lung sliding, a basic sign or normality

Features of Lung Sliding

Several points should be detailed, but should not give the erroneous feeling of complexity.

The most basic point is that low-frequency probes are not adequate to study lung sliding. Unfortunately, several institutions already work with echocardiography-Doppler equipment with 2.5-MHz probes. Operators risk being disappointed when placing such probes over the lungs.

It is important to make it clear that lung sliding is a relative movement of the lung toward the chest wall. Lung sliding involves dynamics that stands out clearly against the motionlessness of the structures located immediately above the pleural line. This is important since a diffuse movement is impossible to avoid in a breathing patient. Dyspnea with use of accessory respiratory muscles raises a particular issue (see below).

Lung sliding can also be very hard to detect with filters such as the dynamic noise filter. These filters yield a softened image using a temporal averaging. Therefore, they are like make-up and give flattering images, but also obscure or mask the true content. The operator must know how to work on a rough, unrefined image. Various factors can be taken into account:

- The amplitude of lung sliding normally increases from the apex to the base. Lung sliding is null at the apex, a sort of starting block. It is maximal at the base.

- The pleural line is interrupted by the posterior shadow of the ribs. If the probe is applied over the costal cartilages, there is no interruption since cartilage does not stop the ultrasound beam.
- Lung sliding is present in spontaneous or conventional mechanical ventilation. It is abolished by jet ventilation.
- Lung sliding is visible in young, old, thin or plethoric patients.
- Lung sliding is not abolished by a dyspnea itself, if one excepts pneumothorax, atelectasis or other causes of abolition.
- Lung sliding can be wide or extremely discrete, but it will have the same meaning. One must thus recognize very discrete lung sliding.
- Pleural sequela, a history of pleurectomy, or talc insufflation, can give conserved or abolished lung sliding (we do not have enough data to make a firm conclusion).
- Lung sliding is present in patients with emphysema. Even a giant emphysema bulla does not abolish lung sliding. This may have basic clinical consequences, when a radiograph is not able to differentiate bulla from pneumothorax, for instance.
- Lung sliding is abolished by apnea, as well as any disorder impairing lung expansion (see Chap. 17).

Lung sliding can be hard to detect in the following cases:

- A history of pleurisy. Lung sliding can be abolished.
- Severe acute asthma. Lung expansion is very diminished. One must pay attention to the slightest movement, as pneumothorax is sometimes the cause of an attack in an asthmatic patient.
- Parietal emphysema. It considerably damages the image (but see below).
- Certain causes of dyspnea with use of accessory respiratory muscles. Use of accessory respiratory muscles gives a sliding , superficial to the pleural line, it is true, but this situation can be misleading at the beginning of operator training. Experience will aid in distinguishing both dynamics.
- Inappropriate technique, unsuitable ultrasound device, inadequate smoothing.

A pathological equivalent to lung sliding can also be described in the situation of abolished lung sliding, but with perception of a kind of vibration arising from the pleural line in rhythm with heart beats (Fig. 16.5). This sign is called lung pulse [11].

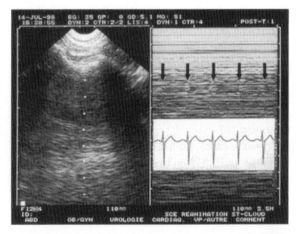

Fig. 16.5. The lung pulse. In this selectively intubated patient, left lung sliding is abolished. Vibrations in rhythm with the heart activity can, however, be recorded at the lung surface, in the time-motion mode (*arrows*)

In the normal subject, the respiration generates lung sliding, and prevents the lung pulse from expressing its presence. In apnea, lung sliding is immediately abolished, and a lung pulse can then be expressed. Apnea is not a stable condition, and the lung pulse is consequently a pathological sign. A lung pulse means that the heart transmits its vibrations through a motionless parenchymatous cushion. In this chapter, devoted to pneumothorax, note that the lung pulse is equivalent to lung sliding. The clinical relevance of this sign, which, in brief, indicates an absence of lung expansion, will be discussed in Chap. 17.

Ultrasound Diagnosis of Pneumothorax

Since air artifacts are the only item investigated here, pneumothorax signs may appear abstract. A rigorous mastery of the signs is required for accurate ultrasound interpretation. Pneumothorax associates abolition of lung sliding, visualization of exclusive A lines, a sign called the lung point, and other signs that are more accessory when the three first signs are present.

Abolition of Lung Sliding

Description

Pneumothorax is a nondependent disorder. It should be sought near the sky, i.e., at the anterior and slightly inferior aspect of the thorax in a supine patient, and with a probe pointing toward

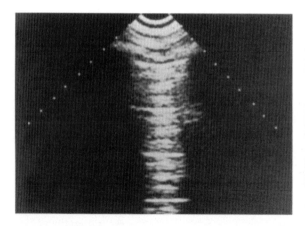

Fig. 16.6. Ultrasound presentation of pneumothorax. The absence of lung sliding cannot be objectified on this single two-dimensional longitudinal scan, but horizontal artifacts arising from the pleural line (three A lines visible here) can be described, and no B line is detected: a pattern called the A-line sign

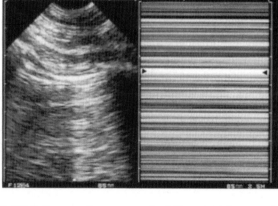

Fig. 16.7. Pneumothorax, sequel of Fig. 16.6. The use of the time-motion mode (*right figure*) objectifies a pattern made of completely horizontal lines, which indicates total absence of motion of the structures located above and below the pleural line (*arrowheads*)

the earth along the earth–sky axis. It has been demonstrated that any free pneumothorax collects at least at the lower half of the anterior chest wall in a supine patient [12]. The first sign of pneumothorax is a complete abolition of lung sliding (Figs. 16.6, 16.7). The pleural line seems to be fixed, a characteristic pattern.

Value of This Sign

A study conducted in a medical ICU compared 43 cases of pneumothorax with 68 normal lungs (on CT). The pathological sign studied was the abolition of nondependent lung sliding. Ultrasound sensitivity was 95% [13]. However, this study classified patients with pneumothorax and parietal emphysema as false-negatives, since lung sliding could not be investigated. On the contrary, if lung sliding cannot be seen, it is not a mistake to consider that lung sliding is absent. By excluding the cases of parietal emphysema, which complicate the methodology, one can assert that, among the feasible cases, ultrasound sensitivity was no longer 95%, but 100%. In other words, all cases of pneumothorax yield abolition of lung sliding. In this study, the main point was a negative predictive value of 100%. Normal lung sliding confidently rules out pneumothorax.

The first description of abolished lung sliding in pneumothorax that we found came from a veterinarian journal [14]. A few studies have also analyzed this pattern [15, 16]. Note that lung sliding is far from summing up the ultrasound signs of pneumothorax. In our first study, ultrasound specificity was only 91% [13], a rate which decreases to 78% when the control population increases [17], and falls to 60% of cases if only ARDS patients are considered (study in progress). The explanation is simple. All controls in our series have benefited from CT (mandatory for ruling out pneumothorax). Hence, a selection bias is created: only patients with an indication for CT, i.e., patients with severe lung disorders, were selected as controls. At the same time, this selection bias is beneficial, since we are in a situation where pneumothorax can be a concern. During the course of ARDS or severe extensive pneumonia, lung sliding is abolished in more than one-third of cases. It is important to state precisely that abolished lung sliding is not specific to pneumothorax. Which disorder can explain an absence of lung sliding? We would say any cause of abolition of lung expansion. Thus, complete atelectasis, but also acute pleural symphysis, or massive lung fibrosis are all factors that may explain abolition of lung sliding.

Analyzing lung sliding makes it possible to recognize a majority of patients as pneumothorax-free. On the other hand, an absence of lung sliding abolition specificity has led us to search for higher-performance signs.

Complete Absence of Lung Rockets: The A-Line Sign

Can artifact analysis be contributive? Definitely yes. An analysis of 41 cases of complete pneumothorax compared with 146 controls studied on CT confirmed that lung rockets were present in 60% of the controls but never in pneumothorax (see Figs. 16.6, 16.7). Absence of lung rockets, in other words, an exclusive A-line profile, what we could call the A-line sign, had a sensitivity of 100% and a specificity of 60% for the diagnosis of pneumothorax [18]. We will see in the next chapter that lung rockets are an ultrasound indicator of interlobular septa thickening, that is, interstitial syndrome [19]. From these notions, we can conclude that lung rockets are generated by the lung itself and never by the parietal pleura. Detecting lung rockets, regardless of the presence or absence of lung sliding, is equivalent to detecting an enabled lung, i.e. also the lung itself.

The low specificity is again explained by the same selection bias as for lung sliding. Precisely, in alveolar-interstitial disorders, lung rockets are massive, wherever the probe is applied (see Chap. 17). The correlation between lung rockets and absence of pneumothorax comes at the right time, since lung rockets are generally present in exactly the cases where lung sliding is abolished (ARDS, extensive pneumonia, etc.).

Association of the both abolition of lung sliding and A-line signs is synergic. Presence of lung sliding or lung rockets identifies a majority of patients who do not have pneumothorax. Specificity of abolished lung sliding and the A-line sign is 96% for the diagnosis of complete pneumothorax [18].

The Lung Point, a Sign Specific to Pneumothorax

We have thus far described signs that were sensitive but not specific. A patient with hard-to-detect or absent lung sliding and absence of interstitial syndrome will have an ultrasound profile of pneumothorax, i.e., a false-positive image. Interestingly, with these signs, we can build a specific sign: with immediate and fleeting visualization at a precise location of the chest wall and along a definite line, at a precise moment of the respiratory cycle, usually inspiration, with the probe strictly motionless, the operator finds either lung sliding, lung rockets, or alteration of A lines, in an area previously observed with no lung sliding and the A-line sign, i.e., patterns that were barely suggestive of pneumothorax (Fig. 16.8). This sign has been called the lung point. When comparing 66 cases of pneumothorax and 233 ICU controls studied on CT, a lung point was observed with a frequency of 66% in the study group and never in the control group, for a sensitivity of 66% and a specificity of 100% for

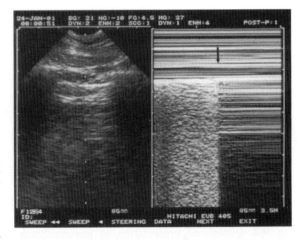

Fig. 16.8. The lung point. In real time (*left*), a transient inspiratory movement is perceived at the pleural line along the middle axillary line, in a patient with pneumothorax of average volume. Time-motion (*right*) shows that the appearance, or here disappearance of lung signs is immediate, according to an all-or-nothing rule (*arrow*)

the diagnosis of pneumothorax [17]. After 7 years of observation, we have never observed a lung point in the countless patients who had no obvious pneumothorax but no need for CT. This sign can be explained if one considers that any lung, at the wall or not, in spontaneous or mechanical ventilation, will slightly increase its volume on inspiration. Therefore, a lung sign will appear at the boundary area, at the precise line where the lung reaches the wall, since the lung surface in contact with the wall will increase (Fig. 16.9). The poor sensitivity of ultrasound is easily explained: major, completely retracted pneumothorax will never touch the wall.

The lung point sign allows each observer to note that lung sliding follows an all-or-nothing rule. It proves that minimal, millimeter-scale pneumothorax will be accurately detected using ultrasound.

Detection of abolished lung sliding with A lines in one area, with lung sliding present or B lines in another area, separated by ribs, for instance, but without lung point is frequent, and it cannot lead to the conclusion of pneumothorax. Focal atelectasis may possibly explain this pattern. Last, in a hasty examination, the liver or the spleen can roughly simulate a lung point.

Other Signs of Pneumothorax

Other signs can sometimes be extremely useful.

For instance, the lung pulse is an equivalent to the normal since its presence rules out pneumo-

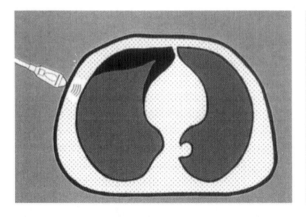 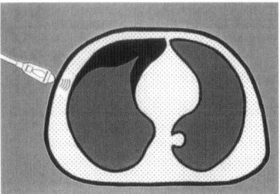

Fig. 16.9. Diagram explaining the origin of the lung point. At the *left*, the probe is applied in front of the pneumothorax, in expiration. At the *right*, after inspiration, the lung has slightly increased its volume, and this is now the lung itself that is located in front of the probe, which remained motionless

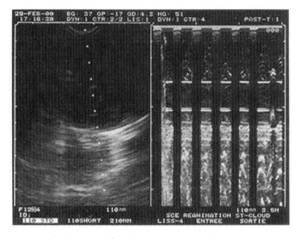

Fig. 16.10. The swirl sign. The *left image* is poorly defined. *Right image*: a rapid succession of air artifacts alternating with transmitted sounds is clearly visible. The rhythm is attributable neither to respiration nor to the heart, but by the swirl of the fluid. Case of hydropneumothorax

thorax. This sign is very frequently observed in critically ill patients, in the numerous cases where abolition of lung sliding is not associated with pneumothorax.

The swirl sign, which has an equivalent at the abdominal level for the diagnosis of occlusion (see p 39), indicates hydropneumothorax. The fluid collection is freely swirled in a depressurized pleural cavity. Consequently, when the probe is applied at bed level and when movements are gently transmitted to the patient, the fluid pleural effusion laps in a highly characteristic manner (Fig. 16.10).

Evaluation of the Size and Location of Pneumothorax

Ultrasound can evaluate the volume of pneumothorax. Radiography offers a very rough indication, since pneumothoraces of any size can be missed [3–7]. In order to optimize ultrasound capabilities, a CT scan should be taken in patients with already proven pneumothorax, i.e., irradiation for scientific reasons but useless for the patient, which raises an ethical issue. Within this limitation, certain entities can be defined. We must first note that so-called minimal pneumothorax, i.e., a thickness of a few millimeters on CT, can be observed over a large area. Therefore, minimal pneumothorax may be observed throughout an extensive area of the anterior chest wall using ultrasound. A study showed that anterior lung point is correlated with minimal and generally radio-occult pneumothorax. Eighty percent of radio-occult pneumothoraces are diagnosed using the lung point [17]. The more lateral the lung point, the more the pneumothorax is substantial. Major pneumothorax yields very posterior or absent lung point.

As regards the usual location of the pneumothoraces encountered in the ICU, the large majority involve at least the anterior zone, especially the lower half in supine patients. It is likely that all life-threatening cases involve this area. In a supine patient, a free pneumothorax collects in the anterior costophrenic sulcus, which is the least dependent area [20]. A study of 56 radio-occult cases of pneumothorax diagnosed on CT confirms this notion: 98% of these pneumothoraces involved at least the lower anterior wall [12]. Only one case of pneumothorax was posterior in this series and seemed not to raise particular concerns.

Practical Detection of Pneumothorax

When pneumothorax is suspected, as detailed in the preceding section, the first step should be to apply the probe at the anterior chest wall (lower half in a supine patient, upper half in a half-sitting patient). Detecting lung sliding, or lung rockets, even if the lung sliding is abolished, rules out complete pneumothorax in a few seconds. If lung sliding is absent and no lung rockets are visible in this area, one should confirm the pneumothorax by detecting a lung point, which will provide information on the volume of the pneumothorax in the same step. If no lung point is detected, it is safer is to use traditional tools such as X-ray or even CT, time permitting. However, if the patient is in critical condition, and if there are clinical signs (suggestive history such as subclavian catheterization, sudden pain, tympanism, abolition of lung sounds), it appears wise to assume that the patient is victim of a genuine pneumothorax and promptly undertake appropriate procedures.

The clinical possibilities are numerous.

1. As regards spontaneous pneumothorax seen in the emergency department, the patient has usually already undergone a chest radiograph performed in the radiology department. This is for the moment necessary, since it would be hard to imagine a medical file without this familiar document. On the other hand, we try to avoid profile incidences, and above all expiratory radiographs at this step. The insertion of the chest tube is planned depending on ultrasound data. First, the insertion site corresponds to an area where the lung is far from the wall (one must know that between a prone and a supine radiograph, the lung takes more lateral room in the supine position). Second, once the tube is fixed, the return of the lung toward the chest wall is checked using ultrasound. Common pneumothoraces return to the wall in 1 or 2 min with aspiration. A time-motion ultrasound view is taken, in order to include a document in the records proving that the pneumothorax was properly treated. No matter where the tube goes, if there is lung sliding, the lung is correctly at the wall. The chest tube is clamped using ultrasound guidance. Persistence of lung sliding indicates that the leakage is sealed. Rapid vanishing of lung sliding means that it can be assumed that the pneumothorax reoccurs after clamping. The tube is eventually withdrawn after the clamping has been judged effective according to the dynamic ultrasound maneuvers. A last ultrasonic view is taken after withdrawal of the tube. To sum up, one should logically find in the patient's records only one radiograph performed at admission (in the absence of presumed pregnancy) and showing the pneumothorax.

2. For pneumothorax occurring under mechanical ventilation, the procedure is the same. However, opportunities to take radiographs are more frequent in ventilated patients. A control radiograph is thus more often available. However, the diagnosis has already been made, before radiography, and the intensivist can prepare the patient while the radiograph is being developed. This procedure saves time and lives. If the patient does not tolerate the pneumothorax, it will not be necessary to wait for the return of the radiograph to treat.

3. In the trauma patient, when the pneumothorax has been proven using ultrasound, the radiograph can be taken if necessary, depending on the severity of the emergency, the patient (pregnant woman, child) and the department's routines. Traumatic pneumothorax should benefit from this approach, which can be achieved in the pre-hospital step. Mastering the ultrasound signs will allow for adequate therapeutic decisions. Note that the blind pleural drainage, which is life-saving only when done in a timely fashion, should disappear from our practice.

4. Routine search after thoracentesis or subclavian catheterization (i.e., when the risk is low but present): a time-motion ultrasound view should replace radiography.

Major Advantages of Ultrasound

The possibility of ultrasound diagnosis of pneumothorax means:

- Positive or negative diagnosis of pneumothorax, at the bedside, i.e., in the emergency situation (respiratory distress, ventilated patient, cardiac arrest, etc.).
- A highly sensitive test: a few millimeters are sufficient to make the diagnosis. The so-called delayed pneumothorax after subclavian catheter

insertion should in fact be recognized immediately. In our opinion, there are no delayed pneumothoraces, there is rather inadequacy of the bedside radiograph.
- Immediate diagnosis, quicker than the quickest bedside radiograph.
- Pre-hospital diagnosis, which is facilitated by the miniature equipment now available.
- No need for lateral decubitus radiographs [9] or transfer to CT.
- A major decrease in irradiation and cost as a consequence of the previous points. All patients, including pregnant women and children, should benefit from this type of diagnosis.

As regards irradiation, it may seem laughable to intellectually and technically invest in lung ultrasonography in order to avoid a few chest radiographs, if a CT (the equivalent of at least 100 or 200 chest radiographs in terms of irradiation) is scheduled for documenting idiopathic pneumothorax. We should then analyze the usefulness of this CT more closely. The main information, a search for contralateral abnormalities, is of little relevance since it has been proven that 89% of patients have such abnormalities, and since CT does not contribute to predicting a new pneumothorax [21].

Limitations and Pitfalls of Ultrasound

Ultrasound fails in the following cases.

Parietal Emphysema

Parietal emphysema is not always associated with pneumothorax. It generates W lines (see Fig. 22.3, p 158) that degrade the signal. However, the pressure of the probe can drive away air and in some cases, lung sliding can be more or less easily identified. Visualization of motionless comet-tail artifacts should be interpreted here with extreme caution at the beginning of training. In fact, the E lines are an apparently dangerous pitfall (Fig. 16.11). A regular layer of air caught between two muscle layers will yield a pattern very similar to lung rockets. It is, however, possible to avoid this pitfall. Any lung ultrasound must begin by the bat sign search. If profuse comet tails hide the ribs, there cannot be lung rockets or any B lines. Similarly, small subcutaneous metallic materials can result in comet tails, distinct from W lines. When an inexperi-

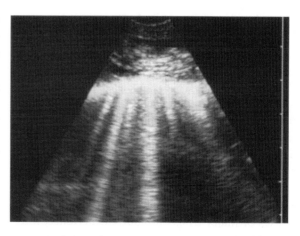

Fig. 16.11. In this longitudinal scan of the chest wall in a traumatized patient with clinical parietal emphysema, well-defined comet-tail artifacts are visible, spreading up to the edge of the screen. They may give the illusion of lung rockets, as in Fig. 16.2, thus ruling out pneumothorax. However, no rib is identified (i.e., the bat sign is absent). The discontinued hyperechoic line from which the comet tails arise is not the pleural line. Layer of parietal emphysema in a patient with massive pneumothorax

enced operator encounters parietal emphysema, traditional tools such as the radiograph or CT, time permitting, are more suitable.

Posterior Locations of Pneumothorax

Although a limitation, posterior locations of pneumothorax are not often of clinical relevance. An ultrasound sign can be expected: abolition of anterior lung sliding. This sign is theoretical but logical, since posterior pneumothorax occurs only if there is large pleural symphysis (see Chap. 17, p 124). We are still awaiting our first case with this probably rare location. Another logical sign will be the absence of posterior lung rockets, a surprising finding after prolonged dorsal decubitus (see next section). As for apical septate pneumothorax, this rare location can logically yield anterior lung rockets with absent lung sliding.

Anterior Septate Pneumothorax

Anterior septate pneumothorax shows large abolition of anterior lung sliding, since the septation assumes large pleural symphysis. Areas of A lines alternate with areas of fixed B lines. This diagnosis is definitely not the easiest.

Imperfect Specificity of Certain Signs

A white radiograph combined with a suggestive ultrasound (abolished lung sliding without lung rockets) renders the diagnosis of pneumothorax probable, but a critically ill patient can also have posterior alveolar consolidation without anterior interstitial syndrome and abolition of lung compliance.

Dyspnea

Cases of major dyspnea require experience, since lung sliding should be distinguished from the muscular sliding generated by accessory respiratory muscles. Note that this concern does not affect spontaneous uncomplicated pneumothorax or pneumothorax occurring in sedated patients. Agitation will render any examination delicate.

Large Dressings

Most dressings prevent ultrasound analysis. One should establish a policy that plans the placement and size of the dressings to keep them to a minimum.

Technical Errors

Using a technique other than the longitudinal technique, focusing on dependent zones, unsuitable filters, an unsteady hand, confusion between B, E and Z lines are all errors that experience eliminates.

In Conclusion

Ultrasound is a seductive answer to a disorder that is very often suggested, less frequently encountered, but which provides potentially awkward problems in the emergency situation. Searching for signs is a simple approach, although rigor is of absolute necessity:

1. Pneumothorax should be searched for using a nondependent approach.
2. The first step is the recognition of the bat sign, which precisely locates the pleural line.
3. A cardiac probe is usually inadequate.
4. Lung sliding is a basic sign of normality, which in practice rules out pneumothorax.
5. Lung sliding can be abolished by pneumothorax, but also atelectasis, acute pleural symphysis, and even apnea.

6. Lung sliding can be erroneously abolished: inadequate frequency, inappropriate filter, or erroneous gain setting.
7. Lung rockets eliminate pneumothorax at the area where they are observed.
8. The lung point is a sign specific to pneumothorax.
9. Ultrasound is superior to bedside chest radiographs for the detection of pneumothorax.

Pneumothorax benefits from fortuitous circumstances that make it especially accessible to ultrasound diagnosis: the anterior area is a highly accessible zone in a supine patient. The harder pneumothorax is to recognize on a radiograph, the easier it is to detect using ultrasound. Severely injured lungs, which are good candidates for barotraumatic pneumothorax, are the very ones in which ultrasound signs will be the most striking.

Finally, since ultrasound holds such an important place, the pertinence of radiological procedures in patients sensitive to irradiations should be questioned.

References

1. Kollef MH (1991) Risk factors for the misdiagnosis of pneumothorax in the intensive care unit. Crit Care Med 19:906–910
2. Pingleton SK, Hall JB, Schmidt GA (1998) Prevention and early detection of complications of critical care In: Hall JB, Schmidt GA, Wood LDH (eds) Principles of critical care, 2nd edn, McGraw Hill, New York, pp 180–184
3. Tocino IM, Miller MH, Fairfax WR (1985) Distribution of pneumothorax in the supine and semirecumbent critically ill adult. Am J Roentgenol 144: 901–905
4. Kurdziel JC, Dondelinger RF, Hemmer M (1987) Radiological management of blunt polytrauma with CT and angiography: an integrated approach. Ann Radiol 30:121–124
5. Hill SL, Edmisten T, Holtzman G, Wright A (1999) The occult pneumothorax: an increasing diagnostic entity in trauma. Am Surg 65:254–258
6. McGonigal MD, Schwab CW, Kauder DR, Miller WT, Grumbach K (1990) Supplemented emergent chest CT in the management of blunt torso trauma. J Trauma 30:1431–1435
7. Gobien RP, Reines HD, Schabel SI (1982) Localized tension pneumothorax: unrecognized form of barotrauma in ARDS. Radiology 142:15–19

8. Steier M, Ching N, Roberts EB, Nealon TF Jr (1974) Pneumothorax complicating continuous ventilatory support. J Thorac Cardiovasc Surg 67:17-23
9. Holzapfel L, Demingeon G, Benarbia S, Carrere-Debat D, Granier P, Schwing D (1990) Diagnostic du pneumothorax chez le malade présentant une insuffisance respiratoire aiguë. Evaluation de l'incidence en décubitus latéral. Réan Soins Intens Med Urg 1:38-41
10. Lichtenstein D (1997) L'échographie pulmonaire: une méthode d'avenir en médecine d'urgence et de réanimation ? (editorial) Rev Pneumol Clin 53: 63-68
11. Lichtenstein D, Lascols N, Prin S, Mezière G (2003) The lung pulse: an early ultrasound sign of complete atelectasis. Intensive Care Med 29:2187-2192
12. Lichtenstein D, Holzapfel L, Frija J (2000) Projection cutanée des pneumothorax et impact sur leur diagnostic échographique. Réan Urg 9 [Suppl 2]: 138
13. Lichtenstein D, Menu Y (1995) A bedside ultrasound sign ruling out pneumothorax in the critically ill: lung sliding. Chest 108:1345-1348
14. Rantanen NW (1986) Diseases of the thorax. Vet Clin North Am 2:49-66
15. Wernecke K, Galanski M, Peters PE, Hansen J (1989) Sonographic diagnosis of pneumothorax. ROFO Fortschr Geb Rontgenstr Nuklearmed150:84-85
16. Targhetta R, Bourgeois JM, Balmes P (1992) Ultrasonographic approach to diagnosing hydropneumothorax. Chest 101:931-934
17. Lichtenstein D, Mezière G, Biderman P, Gepner A (2000) The lung point: an ultrasound sign specific to pneumothorax. Intensive Care Med 26:1434-1440
18. Lichtenstein D, Mezière G, Biderman P, Gepner A (1999) The comet-tail artifact, an ultrasound sign ruling out pneumothorax. Intensive Care Med 25:383-388
19. Lichtenstein D, Mezière G, Biderman P, Gepner A, Barré O (1997) The comet-tail artifact: an ultrasound sign of alveolar-interstitial syndrome. Am J Respir Crit Care Med 156:1640-1646
20. Chiles C, Ravin CE (1986) Radiographic recognition of pneumothorax in the intensive care unit. Crit Care Med 14:677-680
21. Sahn SA, Heffner JE (2000) Spontaneous pneumothorax. New Engl J Med 342:868-874

Chapter 17

Lung

»*The lung is a major hindrance for the use of ultrasound at the thoracic level*«
TR Harrison, Principles of Internal Medicine, 1992, p. 1043

»*Ultrasound imaging is not useful for evaluation of the pulmonary parenchyma*«
TR Harrison, Principles of Internal Medicine, 2001, p 1454

»*Most of the essential ideas in sciences are fundamentally simple and can, in general, be explained in a language which can be understood by everybody*«
Albert Einstein, The evolution of physics, 1937

»*Le poumon…, vous dis-je !*« *(The lung… I tell you!)*
Molière, 1637

In daily practice, examination of the lung can be approached by physical, radiological and CT scan examination. Physical examination is mastered by auscultation, nearly a two- century-old technique [1]. Chest radiography is a century-old technique [2]. CT has been fully available since the 1980s [3]. It is not usual to proceed to lung ultrasonography, since this organ is reputedly inaccessible to this method [4, 5]. Ultrasound artifacts are in principle undesirable structures. Yet the ultrasound representation of the lung is made up solely of artifacts, which can explain this apparently solid dogma (see Figs. 16.1–16.5 and 17.6–17.9). The lung may be an aerated organ, but it is a vital organ.

The ultrasound beam is, it is true, totally stopped when it reaches the lung, or any gas structure. We saw in Chap. 16 that the numerous artifactual signals generated by the gas structures can be described and differentiated from each other. They can be classified into A, B, … Z lines. Indeed, observation shows that the pathological lung basically differs from the normal lung.

One application has already been analyzed, the diagnosis of pneumothorax. It is, in a way, an ultrasound of the »non-lung«. Lung sliding and lung rockets (see Chap. 16) indicate that the very lung surface is visualized.

The Normal Lung Pattern

The lung ultrasound technique was described in Chap. 15 and the normal pattern of the lung in Chap. 16. Let us recall the essential points: the normal lung signal consists of one dynamic sign, lung sliding, and one static sign, the A line, exclusive or predominant.

In diseased lung, virtually any disorder gives a particular signal. Alveolar consolidation, atelectasis, interstitial syndrome, abscess, even pulmonary embolism all have a characteristic pattern.

Alveolar Consolidation

Numerous terms are used in daily practice such as alveolar syndrome, alveolar condensation, density, infiltrate, parenchymatous opacity, pneumonia, bronchopneumonia, pulmonary edema or even atelectasis (a term often misused). This profusion may indicate a certain diagnostic uncertainty. »Hepatization« is an interesting word in the ultrasound field, since the lung and the liver have a similar pattern. The term »alveolar filling« refers to a nonretractile cause. The only and simple term we use is »alveolar consolidation«, since this term does not involve an etiology (infectious, mechanical, hydric).

From the moment the consolidation reaches the visceral pleura, lung consolidation will be perfectly explorable with a short surface probe (Fig. 17.1). The consolidation can be in contact with the pleural line or be visualized through a pleural effusion (see Fig. 15.7, p 99). As early as 1946, Dénier, the father of ultrasound, described this possibility [6]. Ultrasound's potential was defined in the meantime [7–9], but CT correlations are rarely available.

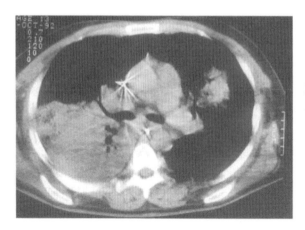

Fig. 17.1. This CT scan of an alveolar consolidation shows a large pleural contact at the posterior aspect of the lung, a condition necessary to make this consolidation accessible to ultrasound. This pleural contact is present in almost all alveolar consolidations seen in acute patients

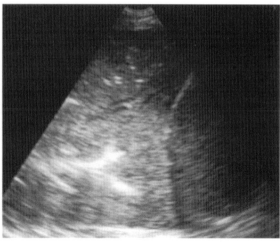

Fig. 17.3. Massive alveolar consolidation of the lower right lobe, longitudinal scan of the lower intercostal spaces. Hyperechoic opacities are visible, punctiform at the *top*, linear at the *bottom*. They indicate air bronchograms

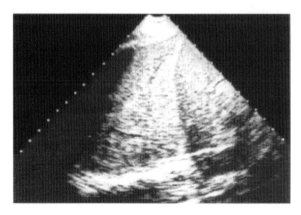

Fig. 17.2. Massive alveolar consolidation of the lower left lobe. The acoustic barrier that is normally expected is replaced with a large tissular suprahrenic mass. This consolidation is substantial. If one takes, in this single scan, a measure in the core–surface axis (vertical on the image), the value is 9 cm. The measure in the horizontal axis of the image, i.e., in the craniocaudal axis, is 8.5 cm here. These dimensions indicate major injury (a consolidation index of 76.5). Note also the homogeneous pattern of the consolidation. Pleural effusion and air bronchogram are not visible. Longitudinal scan of the left base, lateral approach

In our observations, alveolar consolidation yields a pattern characterized by the following items:

1. Tissue pattern. Instead of the usual air barrier, a real image, whose echostructure is a reminder of the hepatic parenchyma, is observed (Fig. 17.2).
2. Boundaries. The superficial boundary is regular, since it is the visceral pleura, i.e., the pleural line in the absence of effusion. The deep boundary can be ragged (the junction between consolidated and aerated parenchyma) or regular, when the whole lobe is involved.
3. Dynamics. The consolidation can have a global dynamics along the craniocaudal axis or no dynamics at all, but no dynamics in the core superficial area as in pleural effusion (see Fig. 15.8, p 99).
4. Echostructure.
4A. Air bronchograms. The consolidation can include numerous punctiform or linear hyperechoic opacities, obviously corresponding to the air bronchograms (Fig. 17.3). These bronchograms, when present, are either dynamic or static:
4A1. The dynamic air bronchogram (Fig. 17.4). Visualization of a dynamics within an air bronchogram has clinical relevance: the air present in the bronchi is subject to a centrifuge inspiratory pressure resulting in its movement toward the periphery. An air bronchogram is thus in continuity with the gas inspired by the patient (either spontaneously or through mechanical ventilation). In other words, a dynamic bronchogram consolidation (DBC) indicates that the consolidation is not retractile: atelectasis

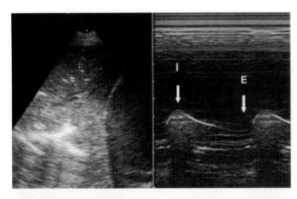

Fig. 17.4. Demonstration of dynamic air bronchogram. The hyperechoic punctiform images, which indicate the air bronchograms within alveolar consolidation (see Fig. 17.2), happen to show an inspiratory centrifuge motion. Time-motion mode perfectly highlights this dynamics (*I*, inspiration. *E*, expiration). This exclusively ultrasonic feature affirms the nonretractile character of this alveolar consolidation

can be ruled out. To detect the dynamic air bronchogram, the bronchus must be in the precise axis of the probe. The operator must avoid confusion with false dynamics such as the out-of-plane effect. This effect will give the erroneous impression that the bronchograms light up: this is a different dynamics.

A consolidation is often associated with an abolition of lung sliding, probably by a decrease in lung expansion. This motionlessness of the lung is a fortuitous condition facilitating the dynamic analysis of its content.

4A2. The static air bronchogram. When no dynamics is observed on an air bronchogram, we speak of static bronchogram consolidation (SBC). This pattern means either that the air bubble is trapped and isolated from the general air circuit (before being dissolved) or that the observation is not correctly located. In the first case, it is tempting to see a sign of atelectasis there, with air still trapped in the bronchi. A study has confirmed that a dynamic air bronchogram was never observed in case of atelectasis, whereas it was observed in 60% of cases of alveolar consolidation of infectious origin [10].

4A3. Consolidation without visible bronchogram. The consolidation can be compact, exclusively tissue-like. We then speak of consolidation with no bronchogram, or NBC (see Fig. 17.2).

4B. Signs of abscess. When the volume of the consolidation is substantial, it is possible to scan this area, in order to check for the homogeneous pattern (air bronchograms excepted). An abscess can then be detected (see »Abscess« p 125).

5. Location of the consolidation. the consolidation can be precisely located, considering the relation with the diaphragm, but also the cutaneous projection. The usual location in a supine, ventilated patient is the lower lobe, i.e., the lower half of the lateral zone, or more posterior. Anterior location is rare, except in complete atelectasis. In case of community-acquired pneumonia, the location can be anywhere. The lower anterior half corresponds to middle-lobe pneumonia. Pneumonia due to pneumococcus usually has extensive contact with the wall, often anterior.

6. Volume. Scanning makes it possible to roughly evaluate the volume of the consolidation. We have found it practical to measure only two dimensions in a single longitudinal scan. For instance, Fig. 17.2 shows a substantial consolidation, with a 90-mm core-to-superficial length, and an 85-mm craniocaudal height.

7. Details. The following signs may or may not have consequences on the etiological diagnosis of the consolidation.
   - The C lines. A real, tissular image touching the surface, with a size on the centimeter scale or less, roughly pyramidal or cupola-shaped (hence the C for cupola), is a small alveolar node, although interstitial disorders (with nodules) may give this pattern (Fig. 17.5).
   - Satellite images. A pleural effusion is often associated with consolidation. When it is not, we speak of dry consolidation.
   The areas near the alveolar consolidation can have an interstitial pattern (with B lines; see below) or a normal pattern, with A lines.
   - The dynamics but also the location of the hemidiaphragm should be described.
   - A deviation of the nearby organs may be informative.

If the definition of the alveolar consolidation includes detection of a tissular pattern, with a reg-

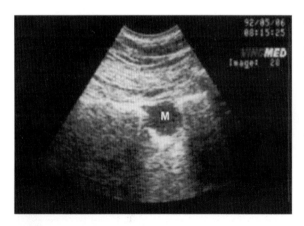

Fig. 17.5. The pleural line is interrupted by a centimeter-scale image, concave in depth (*M*). This is a C line, a sign of very distal alveolar syndrome, or sometimes a nodule

ular superficial boundary and an irregular deep boundary, with craniocaudal or abolished dynamics but without a sinusoid sign, and with more or less hyperechoic punctiform opacities, sensitivity of ultrasound is 90% and specificity 98% when CT is taken as the gold standard [11].

Our search technique varies as a function of the possibility of moving the patient and the therapeutic consequence. In supine patients, stage 2 or stage 3 investigation is usually sufficient (see p 97). Stage 4 is most often carried out in order to make the most exact correlations with CT, but the additional information rarely alters therapeutic plans.

Pitfalls

The distinction between complex pleural effusion and alveolar consolidation is usually easy (see Chap. 15). The sinusoid sign, a deep boundary pattern, air bronchograms, especially when dynamic, are decisive signs. In very rare cases, it is impossible to distinguish the solid part from the fluid part (the ultrasound dark lung; see Chap. 15, p 102).

Abdominal fat should be very similar to alveolar consolidation, but it is a good habit to first locate the hemidiaphragm for easy distinction.

Is such a long description of ultrasound patterns relevant, since radiograph is already available? The answer is yes, above all because alveolar consolidations, especially of the lower lobes, can easily be invisible on bedside radiographs. Second, because ultrasound gives an approach by sections, which allows accurate recognition and measurement of fluid, alveolar syndrome, abscesses, etc.

Acute Interstitial Syndrome

Pleural effusions, pneumothorax and alveolar consolidation are therefore accessible to ultrasound, in spite of the reputation of non-feasibility at the thoracic level. However, the performance of ultrasound does not stop here. Analyzing air artifacts alone, the very ones that supposedly made thoracic ultrasound impossible, make it possible to go further.

Therefore and paradoxically, the detection of an interstitial syndrome is indeed the concern of ultrasound. This application was announced in 1994 [12] and confirmed in 1997 [13]. We will first see how to detect it, then why to detect it.

Acute interstitial syndrome involves a wide range of situations, including adult respiratory distress syndrome, cardiogenic pulmonary edema, bacterial or other pneumonia, chronic interstitial diseases with exacerbation.

The interstitial syndrome is not known to give physical signs, nor is a bedside chest radiograph expected to show interstitial changes, without exception. Even in a good-quality radiograph taken in an ambulatory patient, this diagnosis is particularly difficult, subjective, and a single reader can interpret differently from one day to the next [14].

The Ultrasound Signs

Elementary sign, the comet-tail artifact arising from the pleural line, well defined, erasing A lines, in rhythm with lung sliding and spreading up to the lower edge of the screen without fading, i.e., the ultrasound B line (Fig. 17.6). This description distinguishes the B line from the Z line (Fig. 17.7) and the E line (see Fig. 16.11, p 113).

The elaborated sign is the visualization of several B lines in one longitudinal view between two ribs. This pattern is a reminder of a rocket after lift-off, and is called lung rockets (a practical label). The distance between two B lines at their origin is 7 mm or less. When it is 7 mm, one speaks of B7 rockets (Fig. 17.8). When this distance is less, usually around 3 mm, the B lines are twice as numerous, and we speak of B3 lines or B+ lines (Fig. 17.9).

The pattern that defines interstitial syndrome is the presence of lung rockets wherever the probe is applied at the anterolateral chest wall in a supine or half-sitting patient. The term here is »diffuse rockets«, which implies a bilateral anterior and lateral pattern, from apex to bases. An isolated B line has not yet been shown to be pathological, to our

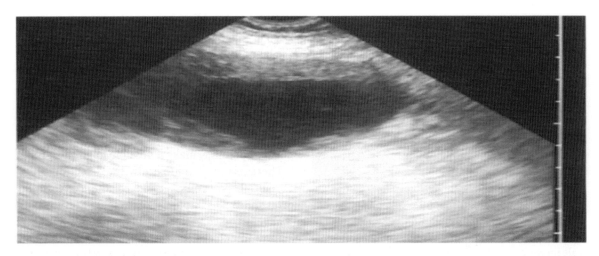

Fig. 17.6. Example of a b line. Arising from the pleural line, an isolated comet-tail artifact, well defined, laser-like, is spreading up to the edge of the screen without fading and erases A lines

Fig. 17.8. Five B lines are identified in this longitudinal scan of the anterior chest wall. They define a pattern reminiscent of a rocket at lift-off. Artifacts are separated from each other by an average distance of 7 mm. Lung rockets are an ultrasound elementary sign of interstitial syndrome

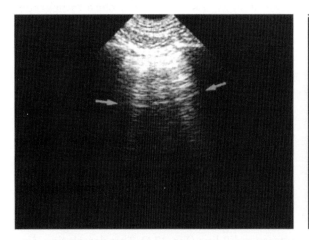
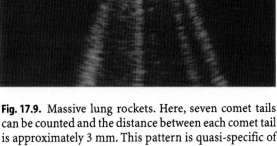

Fig. 17.7. Three vertical, ill-defined artifacts arising from the pleural line and fading after a few centimeters were defined. These artifacts are Z lines, a type of air artifact which should never be confounded with B lines. Because of the clinical importance of this distinction, we prefer to duplicate Fig. 16.3 here. *Arrows:* A line

Fig. 17.9. Massive lung rockets. Here, seven comet tails can be counted and the distance between each comet tail is approximately 3 mm. This pattern is quasi-specific of ground-glass areas. In our experience, this pattern indicates acute interstitial syndrome

knowledge. In order to specify that a B line is isolated, we speak of b line (lower case »b«).

Value of Lung Rocket Signs

In a study including 81 cases of massive alveolar-interstitial syndrome and 119 controls without alveolar or interstitial changes, ultrasound sensitivity based on the previous definition was 92.5% and specificity 94% [13]. Note that feasibility was 100%.

Which structure is at the origin of the comet-tail artifact? Nine items can clearly define it:

1. The comet-tail artifact indicates an anatomical element with a substantial acoustic impedance gradient with the surrounding elements [15], for instance, air and water.
2. The detected element is small, inferior to the resolution power of ultrasound, which is roughly 1 mm, hence not directly visible.
3. This structure is visible at the lung surface.
4. It is visible all over the lung surface.

5. The element is separated from each other by 7 mm.
6. It is present at the last intercostal space in about one-quarter of normal subjects; see Chap. 16.
7. It is correlated with pulmonary edema.
8. It vanishes with the treatment of the pulmonary edema (in a few hours when the edema has cardiogenic origin).
9. It is also present in any interstitial disease.

All these criteria, in a way casting out the nines, are the precise description of thickened interlobular septa. The hypothesis that lung rockets indicate thickened septa has been confirmed: in fact, CT correlations showed that normal structures stop a few centimeters before the lung surface, whereas thickened interlobular septa reach the periphery, i.e., the visceral pleura (Fig. 17.10). In this viewpoint, the ultrasound B lines appear as an ultrasound equivalent of the familiar Kerley's B lines [16]. Note that Kerley's B lines are observed at the bases of 18% of thoracic radiographs of healthy subjects [17]. This number is not very far from the 28% of lung rockets present at the last intercostal space of healthy subjects [13]. The difference probably indicates a slight superiority of ultrasound to detect these very fine elements.

The potential of ultrasound to detect water explains the high performance. Here, water is present in a very small amount, a submillimeter thickness. A thickened interlobular septum is 700 μm thick, versus 300 μm for a normal septum. However, this infinitesimal amount of water is surrounded by air. This mingling is the essential condition required to generate the ultrasound B lines. In addition, clinical observation shows that the interstitial syndrome, especially in pulmonary edema (either cardiogenic or lesional) is a diffuse disorder. This makes its detection immediate wherever the probe is applied. It should be understood that interstitial edema involves all interstitial tissue, the superficial part of it being accessible to ultrasound.

Pathological and Nonpathological Locations of Lung Rockets

- The b lines can be occasionally observed in normal subjects, possibly indicating the small scissura.
- Lung rockets localized at the last intercostal space are found in 28% of normal subjects [13].
- Lung rockets located at the lateral wall but including more than two intercostal spaces

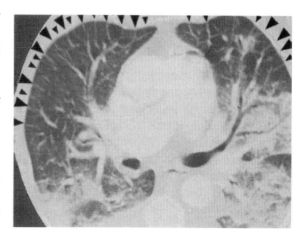

Fig. 17.10. CT scan of massive alveolar-interstitial syndrome. Thickened interlobular septa are visible touching the anterior surface (*arrows*). In a normal subject, no dense structure is visible at the anterior or posterior aspects

above the diaphragm should be considered abnormal. The label used is »extensive lateral rockets.« In general, more posterior analysis usually shows alveolar changes.
- Posterior lung rockets in supine patients are usual, and possibly indicate that the lung water preferentially accumulates in the dependent areas. Analysis of CTs without lung disorders clearly shows these dependent changes. On the other hand, the absence of posterior rockets in a chronically supine patient is singular, and may mean, if validated, substantial hypovolemia.

Clinical Relevance of Lung Rockets

Ultrasound recognition of the interstitial syndrome has several implications, a majority of them already validated.

Ultrasound Diagnosis of Pneumothorax

The recognition of lung rockets immediately rules out complete pneumothorax [18]. Note that this item is basic when lung sliding is very weak or absent, which is a common finding in ARDS. Absence of anterior lung rockets in a patient with a white lung on radiography is suggestive of pneumothorax, but far from specific.

Ultrasound Diagnosis of Pulmonary Edema

Diagnosis Before radiography

In the emergency situation, the physical examination can be atypical in a dyspneic patient with pulmonary edema. We know that interstitial edema precedes alveolar edema [19]. Crackles can be absent at the early stage [20] or be replaced by sibilants in cardiac asthma. Last, fine auscultation can be illusory in a ventilated patient.

In all these cases, ultrasound provides early diagnosis.

Pararadiological Diagnosis

Ultrasound can reinforce the radiograph, once read.

- The chest X-ray, even of good quality, can be difficult to interpret. Let us cite again Fraser, who notes that some radiographs that were interpreted normal on Monday are labeled interstitial on Friday, and by the same reader [14].
- The radiograph can be taken too early. A good-quality radiograph, when taken too early, can be subnormal, even in genuine, very severe pulmonary edemas [21, 22]. The radiograph should clear evidence of advanced stages of edema.
- The radiography can be ill-defined. This is the usual case in emergency. The radiograph is known not to be accurate enough to detect signs of left heart dysfunction. X-ray sensitivity in detecting interstitial edema can range between 18% and 45% [23]. Bedside chest radiography is known to be insufficient for the diagnosis of interstitial syndrome [24]. In addition, Kerley B lines have been described in pulmonary edema and exacerbation of COPD [25].

Nonradiological Diagnosis

When the radiography is not readily available such as in pre-hospital medicine, or, in rare instances, in the hospital itself, or when radiography is not indicated such as in pregnant women or children, and possibly in each patient, ultrasound can find a place.

Differential Diagnosis Between Cardiogenic Pulmonary Edema and Exacerbation of Chronic Obstructive Pulmonary Disease

Presence or absence of lung rockets generally places a dyspneic patient immediately into one of these two groups: diffuse interstitial syndrome or absence of interstitial syndrome. Diffuse bilateral lung rockets is a pattern seen in 100% of cases in cardiogenic acute pulmonary edema vs 8% of cases in patients with exacerbation of COPD [26].

Differential Diagnosis Between Lesional and Cardiogenic Pulmonary Edema

Determining the lesional or cardiogenic origin of a white lung is a frequent task. To oversimplify, water in cardiogenic pulmonary edema is submitted to hydrostatic pressure and moves up to the nondependent areas. In lesional edema, water passively descends to the dependent areas. These movements will have a sonographic outcome: the absence of diffuse anterior lung rockets when there are white lungs on the radiograph are highly suggestive of lesional edema (study in progress).

Diagnosis of Pulmonary Embolism

We will see in a dedicated section that visualizing lung rockets is highly uncommon in this disorder.

Qualitative Estimation of Wedge Pressure

We will not debate on whether wedge pressure provides pertinent or totally outdated information. Some turn their back on this information judged obsolete. The reader can refer to p 180 in Chap. 28, where the problem is detailed more extensively. Our wish is to provide noninvasive data that correlated with wedge pressure for the intensivist who can find such a parameter useful.

Observation shows that the absence of lung rockets is clearly correlated with low wedge pressure. This relies on elementary logic. The same logic indicates that lung rockets are a reflection of lung water. Note that neither right-heart catheterization nor the transesophageal echocardiography provide direct representation of the lung water. Lung rockets are indeed a tracer that directly indicates edematous septal engorgement. In this application, lung ultrasonography will have the advantage of exploring the primary cause of the pulmonary edema, which is as a rule radio-occult. Of course, septa can be thickened by inflammation, and the relation between lung rockets and high wedge pressure is less correlated.

Monitoring Fluid Therapy

The analysis of lung rockets may have an apparently unexpected relevance directly derived from the previous wedge pressure. First observations show that the appearance of lung rockets during fluid therapy is the first change, which occurs before any others (crackles, desaturation or radiographic changes). This is logical since gas exchanges occur at the fine, not yet edematous area of the alveolocapillary membrane [27]. Surface lung ultrasonography will indicate that the septa are dry, and that a safety margin exists if fluid therapy is envisaged. We should remember that the radiological signs of interstitial change precede the clinical signs of pulmonary edema [28].

Evaluation of Lung Expansion

The movement of the pathological comet-tail artifacts can be analyzed and measured. This can give an accurate index of the lung expansion and can have clinical implications. The normal lung excursion is 20 mm at the bases in ventilated patients. It can be completely abolished in pathological conditions.

Monitoring the Ventilatory Parameters in ARDS

According to recent studies of ARDS patients with diffuse attenuations on CT, a positive end-expiratory pressure can induce alveolar recruitment without overdistension, whereas in lobar patients, alveolar recruitment is modest and overdistension of previously aerated areas occurs [29]. A relationship can be established between overdistension and lung rockets. In ARDS, the anterior pattern can display lung rockets or A-line areas. B+ lines are correlated to ground-glass areas [13]. This notion can be of interest for the intensivist who alters the management of the patient as a function of the presence or absence of ground-glass areas (study in progress).

Diagnosis of Nonaerated Lung

The detection of lung rockets in a posterior approach of a supine patient is equivalent to ruling out alveolar consolidation, since an overwhelming majority of cases of alveolar consolidation reach the posterior pleura. In these cases, the posterior aspect of the lung is interstitial, but not alveolar. We previously stated that posterior lung rockets are quasi-physiological in chronically supine patients. Following this logic, if alveolar consolidation is detected in a dependent area, pleural effusion can be ruled out as well.

Atelectasis

Ultrasound patterns in atelectasis have not been extensively described. Artifacts and real image analysis is, however, possible. A number of observations can describe several aspects:

- An immediately available and reliable pattern is the lung pulse. This sign was described in Chap. 16 (see Fig. 16.5, p 108). The lung pulse, which in addition rules out pneumothorax, can be observed within the first seconds of complete atelectasis. A characteristic example is realized in case of selective intubation. Selective intubation creates a sudden and complete left atelectasis. The left lung is aerated, and remains thus a certain time, if an early radiograph is performed. Paradoxically, the lung pulse ultrasound sign is immediately present in 90% of cases [30]. A lung pulse can be visible or invisible, but the abolition of lung sliding is constant, since it is observed in 100% of cases. In addition, the left hemidiaphragm descent is abolished.

Eventually, the lung empties of its gas, and the atelectasis becomes patent, i.e., visible on radiographs. The consolidated lung is thus directly analyzable using ultrasound (Fig. 17.11).

Lung sliding is always abolished in complete atelectasis.

The lung has a tissular pattern. Air bronchograms can most often be observed, but only static air bronchograms should be observed [10]. The absence of any air bronchogram is a very indirect sign of atelectasis.

Fluid bronchograms have been described [31]. They would yield small anechoic tubular structures and be observed in obstructive pneumonia only. We were not able to observe them, or to distinguish them from visible vessels, with our 5-MHz probe.

Very characteristic signs of complete atelectasis are all the signs indicating a loss of lung volume. The intercostal spaces are narrowed. The hemidiaphragm is heightened above the mammary line. The spleen or liver have a frank thoracic location. The mediastinal attraction is one of the more

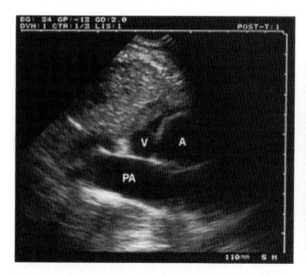

Fig. 17.11. Massive atelectasis of the right lung. Transversal scan of the right anterior third intercostal space. Instead of an acoustic barrier, a tissular image is visible. It shows complete consolidation of the upper right lobe. We can observe the ascending aorta (*A*), the superior vena cava (*V*) and the right pulmonary artery (*PA*), in brief, the mediastinum, which is here frankly shifted to the right. Other pathological points were noted in this ventilated patient: static air bronchograms, phrenic elevation, abolished lung sliding, and lung pulse among others

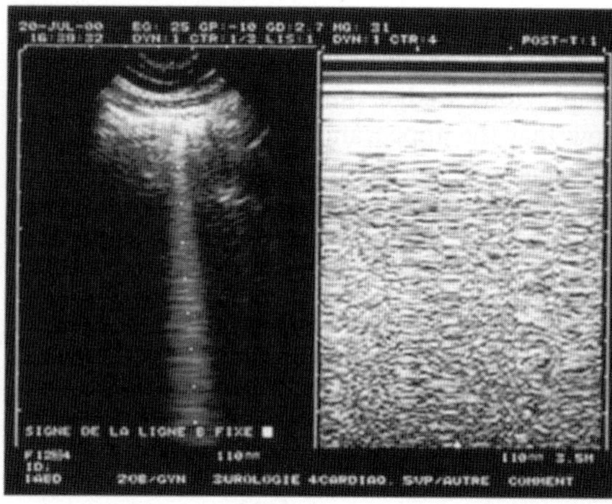

Fig. 17.12. The b line of the left image is completely motionless. A time-motion view at the exact level of this b line objectifies the disorder. A mobile b line would escape at regular intervals outside the cursor line like a pendulum, and would yield a succession of clear and dark bands, and not this homogeneous clear pattern (*right image*). This pattern indicates abolition of the lung expansion

striking patterns (Fig. 17.11). The mediastinum, usually difficult to access, is perfectly analyzable, as during transesophageal examinations. This serendipitous effect allows a clear analysis of usually hidden structures: the vena cava superior at the right (see Fig. 12.20, p 80), the pulmonary artery and its left and right branches, the pulmonary veins, and possibly the main bronchi can be analyzed. Before the treatment of an atelectasis, scanning the mediastinum is recommended. If time lacks, it is always possible to quickly record the data on videotape, and quietly visualize the images later, searching for venous or arterial thromboses, mediastinal tumors, etc.

Acute Pleural Symphysis

Using lung sliding and the comet-tail artifact has allowed us to identify a frequent situation occurring in severe disorders: abolition of lung sliding without pneumothorax (Fig. 17.12). This situation is particularly frequent in ARDS and massive pneumoniae, especially those due to pneumococcus. Patients are generally on mechanical ventilation. In a few cases we could check, inflammatory adhesions of the lung stuck the visceral pleura against the parietal pleura. It is important to know acute pleural symphysis is possible in order not to speak of pneumothorax in these cases. As a rule, lung rockets or a lung pulse will often be present here and thereby rule out pneumothorax.

The diagnostic relevance of this disorder may be to provide an argument to differentiate lesional from cardiogenic pulmonary edema. In cardiogenic edema, only water transudates from the pleura, which cannot impair lung sliding. In lesional edema, there is exudation of fibrin, which may result in the pleural layers sticking.

As regards therapeutic relevance, for the moment, one can only assume that acute pleural symphysis will result in acute restrictive ventilatory disorder. The appropriate therapy is another matter. Note finally that other conditions can abolish lung sliding: complete atelectasis or again pulmonary fibrosis.

Pulmonary Abscess

This disorder is also explored successfully using CT. Bedside radiographs are usually inadequate, since the air–fluid level is not aligned by the X-rays. Ultrasound can be tried as often as necessary. In fact, abscesses are most often peripheral and benefit from a parenchymatous acoustic window.

An abscess within alveolar consolidation appears as a hypoechoic, clearly defined, rather regular image (Fig. 17.13). A collection of gas gives strong echoes.

The air–fluid level is accessible to ultrasound if the probe is applied at the patient's back and points frankly toward the sky. The screen will successively display a fluid image then an air acoustic barrier with a lapping boundary. This assumes, however, a levitation maneuver, which is in practice difficult to achieve. A more accessible maneuver is possible (Fig. 17.14): the approach is as posterior as possible, but without levitation maneuvers. Small bumps are made in the bed. The fluid level will thus be moved. As for bowel occlusion or hydropneumothorax (see pp 39 and 111), the ultrasound translation will be strong, highly suggestive dynamics: one can imagine the dynamics of a shaken glass of water. This sign is called the sign of the air–fluid level, or better, the swirl sign.

Pulmonary Embolism

Sometimes an obvious diagnosis, sometimes tricky, pulmonary embolism remains a daily concern. The abundance of protocols and algorithms indicates that progress is indeed needed. Any help should be studied with attention, especially if non-invasive.

Cardiac and venous signs are detailed in the corresponding chapters. Briefly, a dilatation of the right ventricle is one of several signs, but right heart analysis is frequently difficult in the emergency room using surface ultrasound. Chapter 28 shows that this is not a hindrance. Venous signs are found in a majority of cases (more than 80%) if one makes the effort to search for them in the lower but also upper extremities in ICU patients.

At the lung itself, three signs can be described.

1. The most striking sign in our experience is the presence of a majority of A lines at the anterolateral chest wall. Absence of lung rockets was in fact noted in 91% of 33 cases of severe pulmonary embolism [32]. This pattern is immediately suggestive in a patient without chronic lung disease (asthma, COPD) with sudden dyspnea. This should not be surprising, as it is an

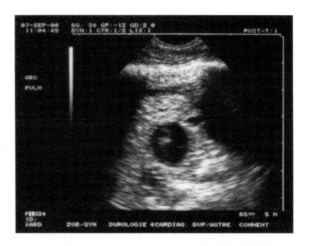

Fig. 17.13. Within an alveolar consolidation, a hypoechoic rounded image is visible in this longitudinal supraphrenic image of the right base. This lung abscess is visible from the echoic surroundings. Located 20 mm below the pleural line (and 30 mm beneath the skin), this abscess is ready for ultrasound-guided aspiration

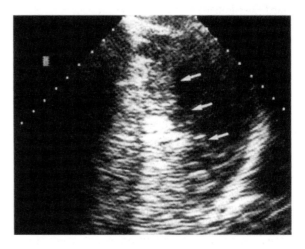

Fig. 17.14. This figure is composed of two zones: one fluid at the *right*, one aerated at the *left*. A roughly horizontal line is thus created (*arrows*). Real-time would show air–fluid swirls. In order to pick up this interface, the ultrasound beam must first enter the fluid zone then the air zone. Obviously, the probe should point to the sky. CT showed a voluminous fluid–air collection in a consolidated lower lobe

ultrasound equivalent of the usually normal radiography. The advantage is immediate, bedside availability. Note that if B3 lines are taken into account, the detection of B3 lines has a negative predictive value of 100% for the diagnosis of severe pulmonary embolism; we are still awaiting our first case.

2. A small, usually radio-occult pleural effusion can be contributive.

3. Some authors describe small peripheral alveolar images as indicating pulmonary infarction [33]. We have seen these patterns (see Fig. 17.5) in less than 4% of cases in a personal series of severe pulmonary embolism. Our explanation is that distal small alveolar infarction may indicate mild pulmonary embolism, a logical deduction, since the smaller the embolism, the more distal the disorder. C lines, as we have called this pattern for years, are more often observed in severe pneumonia with hematogenous extension in our experience.

Routine use of lung ultrasound should rid the intensivist of an old problem: should this patient be transported to the CT room, or worse, to conventional angiography? Should that patient in shock be submitted to the risk of heparin therapy or blind thrombolysis?

Phrenic Disorders

The diaphragm can be dealt with here, since it participates in lung function. The diaphragm has been described in Chaps. 4 (see Fig. 4.9, p 22) and 15 (see Figs. 15.5, p 98 and 15.7, p 99).

The normal inspiratory amplitude of the diaphragm can be analyzed in a longitudinal scan of the liver or the spleen. In spontaneous ventilation in a normal subject, or in conventional mechanical ventilation in a patient without respiratory disorder, it is located between 10 and 15, sometimes 20 mm. Note that a pleural effusion, even substantial, does not affect this amplitude even in mechanical ventilation.

An amplitude under 10 mm, approximately 5 mm, is pathological. Several factors possibly explain a small or abolished phrenic amplitude: pleural symphysis, atelectasis, low tidal volume, or abdominal hyperpressure.

Ultrasound can recognize a phrenic paralysis, a frequent complication in cardiac or thoracic surgery. Transporting the patient to the radioscopy room can be avoided. Phrenic paralysis yields these signs:

- Abnormal movement of the hemidiaphragm in spontaneous ventilation: limited (less than 5 mm) or absent amplitude, or paradoxical movement
- High, intrathoracic location of liver or spleen
- Very diminished or abolished lung sliding
- Absence of thickening of the hemidiaphragm during inspiration, a subtle sign, not always easy to see, and still theoretical (Fig. 17.15).

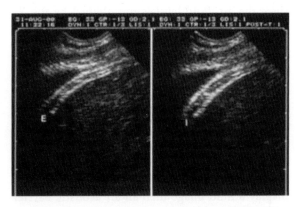

Fig. 17.15. Phrenic respiration. These views objectify the inspiratory thickening of the cupola, increasing from 4 to 6 mm. *E*, expiration. *I*, inspiration

Let us recall that the location of the hemidiaphragm is a first step in any pleural or lung sonography.

Ultrasound in the Etiological Diagnosis of an Alveolar-Interstitial Disorder

The ultrasound pattern alone can suggest certain etiologies.

- Massive alveolar consolidation with dynamic air bronchogram. In our experience, pneumonia caused by pneumococcus yields a pattern of massive alveolar consolidation, with a systematized location: the alveolar consolidation appears all of a sudden, replacing aerated patterns.
- Symmetric pattern associating mild dependent consolidation, mild pleural effusion, diffuse B+ lung rockets. This symmetric pattern is the usual association in acute cardiogenic pulmonary edema.
- Alveolar consolidation without lung rockets. Alveolar syndrome not associated with interstitial syndrome is frequently observed in aspiration pneumonia. This is a logical finding, since aspiration pneumonia is a situation where alveolar lesions occur before interstitial lesions.
- Substantial alveolar consolidation without pleural effusion. A massive consolidation without the smallest pleural effusion probably has a precise meaning. The pneumococcus seems to be associated with this pattern.
- Exclusive diffuse lung rockets, no consolidation, no pleural effusion. This particular profile has

been observed in miliary tuberculosis as well as in pneumocystosis. Lung rockets are usually B3 type, and present at the antero-latero-posterior walls.

Interventional Ultrasound

Some studies have dealt with the possibility of parenchymal puncture for bacteriological investigation purposes. Indeed, procedures were said to be blind, since fluoroscopy was the only means of locating the puncture [34]. If an area of alveolar consolidation is recognized using ultrasound, the precise area where the puncture should be done will be indicated, at the bedside.

Since lung taps are made without ultrasound, let us examine the advantages that ultrasound would provide. First, lung sliding can be assessed. If the lung is adhering to the wall, by pleural symphysis, for instance, the risk of pneumothorax will theoretically be limited. This is the case for fixed alveolar consolidations, i.e., consolidations associated with abolished lung sliding. Second, if the needle crosses completely consolidated lung without contact with the airways, the risk of pneumothorax will clearly fall. Third, if the puncture is posterior, the heavy lung will weigh its full weight over the hole. Using all these precautions, it should be interesting to compare the risk of pneumothorax with the rate of pneumothorax that occurs under a plugged telescopic catheter, an infrequent but possible complication. Fourth, the route is very direct: bacteria swarm just a few centimeters deep below the skin, making the risk of contamination very low. A plugged telescopic catheter will take a very long route: risk of contamination is a main concern. The ultrasound approach has the advantage of an extremely simple procedure, when compared to the invasive ones (fiberscope, plugged telescopic catheter).

Technically, a fine 21-gauge needle should be used. A substantial vacuum will be needed in order to obtain a small drop of brown material. The best results in our institution are obtained when the syringe is directly sent to the laboratory, with the needle inserted, and without additional fluid (serum or other).

Community-acquired as well as nosocomial pneumonia could be approached using the following criteria: large pleural contact allowing good ultrasound location, substantial consolidation, and abolition of lung sliding. These criteria are often present.

Large series will specify more fully where this procedure should be placed. The other invasive or semi-invasive procedures involve a small but present risk, accuracy rates far from perfect and other drawbacks (e.g., cost). Our series show a rate of positive bacteriology of 50%. When positive, the microbes usually swarm in the specimen sent to the laboratory. With the previously described criteria being present, pneumothorax never occurred as a consequence of the procedure.

As regards pulmonary abscesses, ultrasound-guided aspiration was described [35]. This allows direct bacteriological diagnosis. Pleural symphysis and the surrounding alveolar consolidation theoretically protects from risks of pneumothorax. Ultrasound-guided aspiration is not yet proposed in first line treatment, but should be reserved for resistant or severe abscesses: those that are seen, by definition, in the ICU.

References

1. Laënnec RTH (1819) Traité de l'auscultation médiate, ou traité du diagnostic des maladies des poumons et du cœur. J.A. Brosson & J.S. Chaudé, Paris
2. Williams FH (1986) A method for more fully determining the outline of the heart by means of the fluoroscope together with other uses of this instrument in medicine. Boston Med Surg J 135: 335–337
3. Hounsfield GN (1973) Computerized transverse axial scanning. Br J Radiol 46:1016–1022
4. Friedman PJ (1992) Diagnostic tests in respiratory diseases. In: Harrison TR (ed) Harrison's principles of internal medicine. 12th edn. McGraw-Hill, New York, p 104
5. Weinberger SE, Drazen JM (2001) Diagnostic tests in respiratory diseases. In: Harrison TR (ed) Harrison's principles of internal medicine, 14th edn. McGraw-Hill, New York, pp 1453–1456
6. Dénier A (1946) Les ultrasons, leur application au diagnostic. Presse Méd 22:307–308
7. Weinberg B, Diakoumakis EE, Kass EG, Seife B, Zvi ZB (1986) The air bronchogram: sonographic demonstration. Am J Roentgenol 147:593–595
8. Dorne HL (1986) Differentiation of pulmonary parenchymal consolidation from pleural disease using the sonographic fluid bronchogram. Radiology 158:41–42
9. Targhetta R, Chavagneux R, Bourgeois JM, Dauzat M, Balmes P, Pourcelot L (1992) Sonographic approach to diagnosing pulmonary consolidation. J Ultrasound Med 11:667–672
10. Lichtenstein D, Mezière G, Seitz G (2002) Le »bronchogramme aérien dynamique«, un signe échogra-

phique de consolidation alvéolaire non rétractile. Réanimation 11 [Suppl]3:98
11. Lichtenstein D, Lascols N, Mezière G, Gepner A (2004) Ultrasound diagnosis of alveolar consolidation in the critically ill. Intensive Care Med 30: 276-281
12. Lichtenstein D (1994) Diagnostic échographique de l'œdème pulmonaire. Rev Im Med 6:561-562
13. Lichtenstein D, Mezière G, Biderman P, Gepner A, Barré O (1997) The comet-tail artifact: an ultrasound sign of alveolar-interstitial syndrome. Am J Respir Crit Care Med 156:1640-1646
14. Fraser RG, Paré JA (1988) Diagnoses of disease of the chest, 3rd edn. WB Saunders Company, Philadelphia
15. Ziskin MC, Thickman DI, Goldenberg NJ, Lapayowker MS, Becker JM (1982) The comet-tail artifact. J Ultrasound Med 1:1-7
16. Kerley P (1933) Radiology in heart disease. Br Med J 2:594
17. Felson B (1973) Interstitial syndrome. In: Felson B (ed) Chest roentgenology, 1st edn. WB Saunders, Philadelphia, pp 244-245
18. Lichtenstein D, Mezière G, Biderman P, Gepner A (1999) The comet-tail artifact, an ultrasound sign ruling out pneumothorax. Intensive Care Med 25: 383-388
19. Staub NC (1974) Pulmonary edema. Physiol Rev 54:678-811
20. Braunwald E (1984) Heart disease. W.B. Saunders, Philadelphia
21. Stapczynski JS (1992) Congestive heart failure and pulmonary edema. In: Tintinalli JE, Krome RL, Ruiz E (eds) Emergency medicine: a comprehensive study guide. Mc Graw-Hill, New York, pp 216-219
22. Bedock B, Fraisse F, Marcon JL, Jay S, Blanc PL (1995) Œdème aigu du poumon cardiogénique aux urgences: analyse critique des éléments diagnostiques et d'orientation. Actualités en réanimation et urgences. In: Actualités en réanimation et urgences. Arnette, Paris, pp 419-448
23. Badgett RG, Mulrow CD, Otto PM, Ramirez G (1996) How well can the chest radiograph diagnose left ventricular dysfunction? J Gen Intern Med 11:625-634
24. Rigler LG (1950) Roentgen examination of the chest: its limitation in the diagnosis of disease. JAMA 142:773-777
25. Costanso WE, Fein SA (1988) The role of the chest X-ray in the evaluation of chronic severe heart failure: things are not always as they appear. Clin Cardiol 11:486-488
26. Lichtenstein D, Mezière G (1998). A lung ultrasound sign allowing bedside distinction between pulmonary edema and COPD: the comet-tail artifact. Intensive Care Med 24:1331-1334
27. Rémy-Jardin M, Rémy J (1995) Maladies pulmonaires infiltrantes diffuses à traduction septale exclusive ou prédominante. In: Rémy-Jardin M (ed) Imagerie nouvelle de la pathologie thoracique quotidienne. Springer-Verlag, Paris, p 123-154
28. Chait A, Cohen HE, Meltzer LE, VanDurme JP (1972) The bedside chest radiograph in the evaluation of incipient heart failure. Radiology 105:563-566
29. Puybasset L, Gusman P, Muller JC, Cluzel P, Coriat P, Rouby JJ and the CT Scan ARDS Study Group (2000) Regional distribution of gas and tissue in acute respiratory distress syndrome. III. Consequences for the effects of positive end-expiratory pressure. Intensive Care Med 26:1215-1227
30. Lichtenstein D, Lascols N, Prin S, Mezière G (2003) The lung pulse, an early ultrasound sign of complete atelectasis. Intensive Care Med 29:2187-2192
31. Yang PC, Luh KT, Chang DB, Yu CJ, Kuo SH, Wu HD (1992) Ultrasonographic evaluation of pulmonary consolidation. Am Rev Respir Dis 146:757-762
32. Lichtenstein D, Loubière Y (2003). Lung ultrasonography in pulmonary embolism. Chest 123:2154
33. Mathis G, Dirschmid K (1993) Pulmonary infarction: sonographic appearance with pathologic correlation. Eur J Radiol 17:170-174
34. Torres A, Jimenez P, Puig de la Bellacasa JP, Celis R, Gonzales J, Gea J (1990) Diagnostic value of nonfluoroscopic percutaneous lung needle aspiration in patients with pneumonia. Chest 98:840-844
35. Yang PC, Luh KT, Lee YC, Chang DB, Yu CJ, Wu HD, Lee LN, Kuo SH (1991) Lung abscesses: ultrasound examination and ultrasound-guided transthoracic aspiration. Radiology 180:171-175

Lung Ultrasound Applications

Now that we are more familiar with lung signs, the present chapter presents some of the clinical potentials.

Why Such a Delay for Lung Ultrasound to Become Popular?

Considering the numerous applications seen in the preceding chapters, we can wonder why lung ultrasound took so many years to develop. When the present lines were written, the lung was rarely present in the ultrasound or intensive care textbooks, and lung ultrasound was even less a part of emergency procedures. One explanation is that basic applications such as pneumothorax, pneumonia, interstitial syndrome, atelectasis, etc., are ignored. A dogma condemning lung ultrasound until now is partly responsible for this situation. Another possible explanation is that the radiologist, who usually handles ultrasound, has easy access to CT or MRI. CT answers, it is true, a majority of critical questions at the thoracic level. One would thus have passed directly from the radiographic era to the scanographic era. CT developed just after ultrasound, and has, in a way, buried it alive. The problem is completely different for the intensivist, who must answer vital questions in real-time, and for the physician who wants to limit irradiation. Ultrasound's use in examining the lung is judged suboptimal by some authors [1], with whom we obviously agree [2].

For instance, it is striking to see that thoracic ultrasonography is limited to the sole diagnosis of fluid pleural effusion in general reviews [3–5]. Yet today this application is still sometimes forgotten in recent reviews [6]. Practically speaking, the alternative is bedside radiography or CT [7].

The Seven Principles of Lung Ultrasound

As seen in the preceding chapters, a both accurate and reliable collection of signs exists at the lung level, based on seven main principles.

1. The thorax is an area where air and water are intimately mingled. Air rises, water descends. It is thus basic to define dependent disorders and nondependent disorders, to specify the patient's position and the area where the probe is applied.
2. The lung surface is extensive. This is the largest organ in the human body and its surface can be divided into well-defined areas (see Chap. 15).
3. All lung signs arise from the pleural line.
4. Lung signs are mainly based on the analysis of the artifacts, which are usually undesirable structures.
5. The signs are generally dynamic.
6. Nearly all acute disorders of the thorax (pneumothorax, pleural effusions, a majority of alveolar consolidations, interstitial syndrome) come in contact with the surface. This explains the potential of lung ultrasound, paradoxical only at first view. The potential of ultrasound to diagnose these disorders stems most particularly from its capability to clearly distinguish air and water. In addition, a high feasibility, between 98% and 100% [8–12] can be explained by the superficial state of the lung. The examination will therefore be made with optimism in any patient.
7. Last, a simple, two-dimensional apparatus meets the optimal criteria for this task.

The lung, a vital organ, becoming accessible to ultrasound using a simple technique, is not only progress in imaging. It is above all a step toward the concept of the ultrasonic stethoscope.

One Way to Approach Lung Ultrasonography

Air creates a complete acoustic barrier. Water is an excellent acoustic transmitter. Between these extreme cases, various degrees of echogenicity are encountered. The data that follow do not correspond to scientific manipulations, but rather to a rough estimation (Table 18.1).

Suggestion for Classifying Air Artifacts

The artifacts used for emergency diagnoses are numerous, and an overview may be useful to clarify things. Figure 18.1 provides this overview.

Lung Ultrasound Versus Radiography and Tomodensitometry in the Intensive Care Unit

It may seem bold to compare ultrasound to chest radiography (which we have done throughout the three previous chapters), and ultimately disrespectful to dare the comparison with tomodensitometry. Yet, if one wishes to obtain useful information rather that a fine image, observation shows that ultrasound can replace almost all the bedside chest radiographs, and a majority of CTs.

Lung Ultrasound and Bedside Radiography

The intensivist knows the inadequacies of the bedside chest radiograph [13–19]. Several basic emergency diagnoses can be occulted: pneumothorax (even tension pneumothorax), pleural effusions (even abundant), alveolar consolidation (mostly of the lower lobes), and interstitial syndrome (a diagnosis that is not required from a bedside radiograph).

In fact, bedside radiography provides information only when the disorders are advanced. Pointing out these drawbacks can be awkward with a procedure as popular and familiar as radiography has been for over a century [20, 21]. Excellent radiologists, it is true, know how to read bedside radiographs, but they are rare, and not available 24 h a day in small, non-university-affiliated hospitals. We strongly believe that the study of ultrasound signs is, paradoxically, much easier to reproduce.

In the case of a radiological white lung, for instance, ultrasound immediately details the fluid and the alveolar components. It will also diagnose occult pneumothorax and phrenic rupture.

Lung Ultrasound and Thoracic Tomodensitometry

The inadequacies of CT are not often highlighted. The community has retained the overwhelming advantage of providing a good overview, an advantage that will certainly not be contested here. Ultrasound must now earn its place facing this heavyweight of imaging. Let us view the CT in the light of seven major concerns:

1. The need for transportation. This is the major drawback in an emergency.
   - The delay from the decision to perform a CT to the moment when the patient can benefit from therapeutic changes subsequent to the CT results remains substantial. This problem is only slightly remedied with the CT units with rapid (one should say pseudo-rapid) acquisition.
   - An unstable patient is at permanent risk.
   - Multiple life-support equipment (catheters, tubes) can be harmed.
   - The intensivist must passively assist the patient during the entire procedure and cannot deal with other emergencies. It should be

Table 18.1. Degree of aeration and ultrasound signs

Degree of aeration	Pathological disorder	Ultrasound pattern
100%	Pneumothorax	A lines and abolished lung sliding
98%	Normal lung	A lines with present lung sliding
95%	Thickening of the interlobular septa	B7 lines
80%	Ground-glass areas	B3 lines
10%	Alveolar consolidation	Hepatization with numerous air bronchograms
5%	Atelectasis	Hepatization with rare or absent air bronchograms
0%	Pleural effusion	Anechoic collection

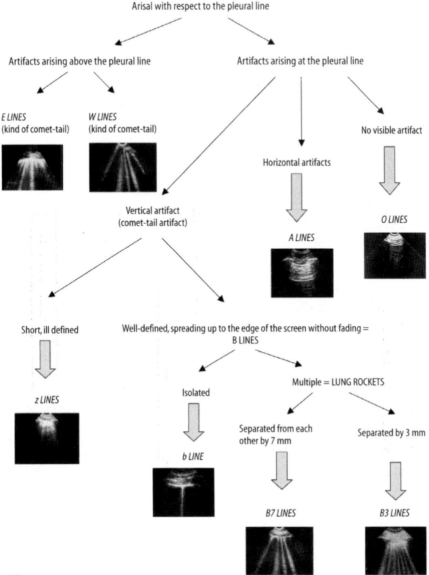

Fig. 18.1. Air artifacts

recalled here that during the night only one intensivist is present for the ICU and all the hospital's extreme emergencies.
- Perfect asepsis is impossible to guarantee in a patient with multiple infections, who therefore becomes a »bacteriological bomb« for the hospital.
- Last, transportation of unstable patients is inevitably a strain for the medical team.
2. Irradiation is substantial. One chest CT scan is 50–200 times more irradiating than a chest radiography. When a CT is performed at the chest level of a woman 30 years of age or under, the risk of having breast cancer is increased by 35% [22]. Deleterious side effects of CT in the child are now acknowledged [23]. Investigation of lung disorders in pregnant women also raises concerns [24].
3. Iodine generates vascular overcharge, risk of anaphylactic shock and renal injury.
4. Diagnostic inadequacies. CT does not resolve all problems. The distinction between alveolar consolidation and pleural effusion can be impossible without iodine injection. Septations within a pleural effusion are not visible. Interstitial syndrome can be hard to detect in ventilated patients. CT will detect an alveolar consolidation, but the dynamic features of the air bronchograms are not detected. Abolition of lung expansion can in no way be documented by a single CT acquisition. Minimal pneumothorax can be missed if images have been acquired

Table 18.2. Performance of ultrasound vs CT in ARDS

Data	Pneumothorax		Pleural effusion		Alveolar consolidation		Interstitial syndrome	
CT	(+)	(−)	(+)	(−)	(+)	(−)	(+)	(−)
Ultrasound (+)	1	0	45	1	55	0	53	2
Ultrasound (−)	0	69	7	17	5	10	2	13
Performance of ultrasound [25]:								
	Sensitivity		Specificity					
Pneumothorax	100%		100%					
Pleural effusion	86%[a]		94%					
Alveolar consolidation	91%		100%					
Interstitial syndrome	96%		86%					

[a] 97% for more than 10 mm maximal thickness effusions.

at inspiration, which is usual with CT. The diaphragm is not well studied by transverse scans, and its dynamics not at all. These points are precisely the strong points of ultrasound. The study of lung signs is a highly dynamic field, and real-time ultrasound is particularly well adapted to it.

5. Intrinsic quality of the image. Daily observations show that CT does not provide optimal-quality images. The signal is impaired by numerous artifacts such as intracavitary devices such as catheters. The arms of the patient cannot always be shifted and are a source of degradation of the image. Respiratory or cardiac dynamics create a blurred pattern. Even when the conditions are optimal, observations show that the focal resolution power of CT is less than that of ultrasound (see Fig. 8.12, p 52). To sum up, when the patient comes back from CT, the additional information sometimes lacks clarity and sharpness.
6. The cost is high when compared to ultrasound, which can be liberally performed without harm, once acquired. Maintenance should be considered for an objective evaluation of costs. For instance, the failure rate caused by CT breakdowns is much higher than that of a basic ultrasound unit, which thus remains an essential tool in emergency diagnosis.
7. Answers to clinical questions. Both ultrasound and CT can provide both qualitative and quantitative answers to basic questions. Table 18.2 shows the performance of ultrasound compared to high-resolution CT in ARDS patients: the performance of ultrasound is not far from 100%. As regards the false-positives and false-negatives of ultrasound, studies in progress may demonstrate that CT can be wrong in some cases, a delicate assertion regarding a gold standard of this magnitude. Yet one situation can be verified immediately: the frequent situation where ultrasound detection of anterior lung rockets is associated with posterior major disorders (alveolar consolidation), but CT shows the posterior alveolar consolidation but no anterior interstitial changes. Should, in these cases, ultrasound be accused of being too sensitive? Or maybe this is an indirect but definite proof that CT is not always able to detect fine interstitial changes.

Ultrasound of Acute Dyspnea

This application, a combination of the others, is described in Chap. 28.

A study was conducted regarding acute dyspnea seen by the intensivist. It showed that the ultrasound data alone provided a correct diagnosis in 85% of the cases, whereas the diagnosis made by the senior physician in the emergency room (or in the pre-hospital instances) with traditional tools (clinical examination, laboratory tests, chest radiograph) was correct in only 52% of the cases [26]. These numbers show that this diagnosis is a particularly difficult one. Ultrasound is here a major tool. A large study will soon confirm our preliminary results.

Conclusions

This chapter could be closed by underlining that ultrasound should not be opposed to radiography or CT, but is rather complementary. However, the distinctive features of ultrasound do indeed set this method apart. Ultrasound can be used at the bedside, with very limited logistics (a simple electrical socket is enough) and at minimal cost, even by the intensivist in order to gain crucial time, with none of the invasiveness discussed in this chapter. This is clearly a tool like no other in the intensivist's armamentarium.

Once these applications are well known, accepted and mastered, lung ultrasound will have a first-line place in intensive care medicine. This method should, with time, progressively diminish the place of bedside radiography and even CT. Of course, a well-trained operator will recognize ultrasound's inadequacies and if necessary immediately use the more conventional CT. It is precisely when these limitations are mastered that ultrasound becomes the high-precision tool it truly is.

References

1. Henneghien C, Remacle P, Bruart J (1986) Intérêt et limites de l'échographie en pneumologie. Rev Pneumol Clin 42:1–7
2. Lichtenstein D (1997) L'échographie pulmonaire: une méthode d'avenir en médecine d'urgence et de réanimation ? (editorial) Rev Pneumol Clin 53: 63–68
3. Mueller NL (1993) Imaging of the pleura, state of the art. Radiology 186:297–309
4. McLoud TC, Flower CDR (1991) Imaging the pleura: sonography, CT and MR imaging. Am J Roentgenol 156:1145–1153
5. Matalon TA, Neiman HL, Mintzer RA (1983) Noncardiac chest sonography, the state of the art. Chest 83:675–678
6. Desai SR, Hansel DM (1997) Lung imaging in the adult respiratory distress syndrome: current practice and new insights. Intensive Care Med 23:7–15
7. Ivatury RR, Sugerman HJ (2000) Chest radiograph or computed tomography in the intensive care unit? Crit Care Med 28:1033–1039
8. Lichtenstein D, Menu Y (1995) A bedside ultrasound sign ruling out pneumothorax in the critically ill: lung sliding. Chest 108:1345–1348
9. Lichtenstein D, Mezière G, Biderman P, Gepner A, Barré O (1997) The comet-tail artifact: an ultrasound sign of alveolar-interstitial syndrome. Am J Respir Crit Care Med 156:1640–1646
10. Lichtenstein D, Mezière G (1998). A lung ultrasound sign allowing bedside distinction between pulmonary edema and COPD: the comet-tail artifact. Intensive Care Med 24:1331–1334
11. Lichtenstein D, Mezière G, Biderman P, Gepner A (1999) The comet-tail artifact, an ultrasound sign ruling out pneumothorax. Intensive Care Med 25: 383–388
12. Lichtenstein D, Mezière G, Biderman P, Gepner A (2000) The lung point: an ultrasound sign specific to pneumothorax. Intensive Care Med 26:1434–1440
13. Greenbaum DM, Marschall KE (1982) The value of routine daily chest X-rays in intubated patients in the medical intensive care unit. Crit Care Med 10:29–30
14. Henschke CI, Pasternack GS, Schroeder S, Hart KK, Herman PG (1983) Bedside chest radiography: diagnostic efficacy. Radiology 149:23–26
15. Janower ML, Jennas-Nocera Z, Mukai J (1984) Utility and efficacy of portable chest radiographs. Am J Roentgenol 142:265–267
16. Peruzzi W, Garner W, Bools J, Rasanen J, Mueller CF, Reilley T (1988) Portable chest roentgenography and CT in critically ill patients. Chest 93:722–726
17. Wiener MD, Garay SM, Leitman BS, Wiener DN, Ravin CE (1991) Imaging of the intensive care unit patient. Clin Chest Med 12:169–198
18. Winer-Muram HT, Rubin SA, Ellis JV, Jennings SG, Arheart KL, Wunderink RG, Leeper KV, Meduri GU (1993) Pneumonia and ARDS in patients receiving mechanical ventilation: diagnostic accuracy of chest radiography. Radiology 188:479–485
19. Tocino IM, Miller MH, Fairfax WR (1985) Distribution of pneumothorax in the supine and semi-recumbent critically ill adult. Am J Roentgenol 144:901–905
20. Roentgen WC (1895) Ueber eine neue Art von Strahlen. Vorläufige Mittheilung, Sitzungsberichte der Wurzburger Physik-mediz Gesellschaft 28:132–141
21. Williams FH (1901) The Roentgen rays in medicine and surgery. MacMillan, New York
22. Hopper KD, King SH, Lobell ME, Tentlave TR, Weaver JS (1997) The breast: in-plane X-ray protection during diagnostic thoracic CT. Radiology 205:853–858
23. Brenner DJ, Elliston CD, Hall EJ, Berdon WE (2001) Estimated risks of radiation-induced fatal cancer from pediatric CT. Am J Roentgenol 176:289–296
24. Felten ML, Mercier FJ, Benhamou D (1999) Development of acute and chronic respiratory diseases during pregnancy. Rev Pneumol Clin 55:325–334
25. Lichtenstein D, Cluzel P, Grenier P, Coriat P, Rouby JJ (1997) Apport de l'échographie pulmonaire dans le S.D.R.A. Réan Urg 6:781
26. Lichtenstein D, Mezière G (2003) Ultrasound diagnosis of an acute dyspnea. Crit Care 7 [Suppl]2:S93

CHAPTER 19

Mediastinum

Can the mediastinum be analyzed within a general ultrasound approach, i.e., using a route other than the transesophageal route? Certainly yes, with an effort to sort out perspective, and if one accepts a low feasibility rate. A small probe will be a precious tool here, as elsewhere. A suprasternal approach has been described [1]. A parasternal approach is contributive, when the mediastinum is shifted to one side. Sometimes, through a not perfectly closed sternotomy, it is again possible to have a modest route for ultrasound.

Thoracic Aorta

In good conditions, which depend to a large extent on the patient's morphotype, it is possible to analyze:

- Initial aorta via the left parasternal route (Fig. 19.1)
- Ascending aorta via the supraclavicular route (Fig. 19.2)

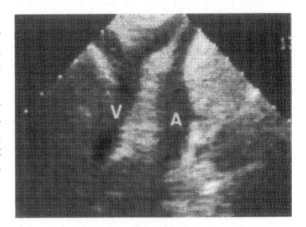

Fig. 19.2. Ascending aorta (*A*), inside the superior vena cava (*V*). Right supraclavicular approach. The origin of the brachiocephalic artery can be seen

- Aortic arch and the three supra-aortic trunks via the suprasternal route (Fig. 19.3)
- Descending thoracic aorta, over several centimeters, behind the heart, via the cardiac apical route (Fig. 19.4)

The abdominal aorta is then followed via the abdominal route up to its bifurcation (Fig. 19.5). It is thus possible to reconstitute a puzzle. The aortic isthmus is, however, generally missing from this puzzle.

A left pleural effusion (for instance, a hemothorax in the case of aneurysm leakage) provides an acoustic window that makes the analysis of the descending aorta possible, via the posterior route (see Fig. 15.14, p 101).

A thoracic aortic aneurysm gives a large mediastinal mass at the aorta. The walls of the aorta generally have a sacciform pattern. The content can show massive thrombosis and then appear as a tissular mass (Fig. 19.6). However, this mass will contain a central lumen, with a stratified periphery. Often, the most central layers of the thrombosis are still mobile, and one can see them driven

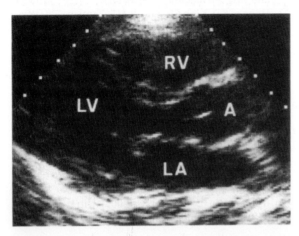

Fig. 19.1. Initial aorta (*A*) visible in a parasternal long-axis scan, between left auricle (*LA*) and right ventricle (*RV*). *LV*, left ventricle. Note that in this scan, the right of the image corresponds to the head of the patient

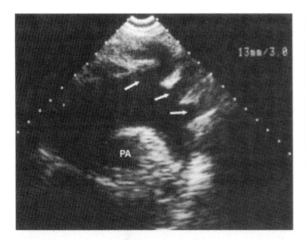

Fig. 19.3. Rare observation of the aortic arch in a young woman with a favorable morphotype, suprasternal approach. The origin of the supra-aortic trunks (*arrows*) and the right pulmonary artery (*PA*) in transverse section are exposed in detail

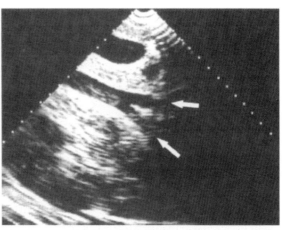

Fig. 19.5. Terminal aorta, sequel of Figs. 19.4 and 4.1. *Arrows*, origin of the iliac arteries. This type of image can replace more invasive modalities such as CT or angiography in emergency situations

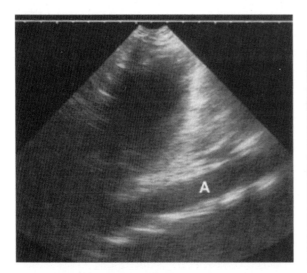

Fig. 19.4. The descending thoracic aorta is exposed over 12 cm in this scan that exploits the cardiac window (apical scan of the heart)

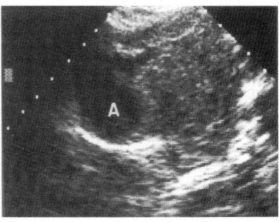

Fig. 19.6. Thoracic aorta aneurysm. Suprasternal scan in a patient in shock with thoracic pain. Note the substantial thrombosis, with regular layers. *A*, circulating lumen of the aorta

back to the periphery a few millimeters in each systole.

In the case of thoracic aortic dissection (Fig. 19.7), an enlarged lumen of the aorta can be observed, and in some cases the intimal flap. This flap has an anatomical shape, i.e., never completely regular, and in our opinion is easily distinguished from the numerous artifacts that are always too regular and generally located in a strictly parallel or meridian plane. However, the search can be difficult, depending on the morphotype, the situation of the flap with respect to the probe axis, and probably also the operator's experience here.

The supra-aortic vessels can be followed to various lengths, but the application seems rare, at least in medical ICU use (see Chap. 21).

Acute Mediastinitis

Studying the mediastinal content after cardiac surgery can be delicate. However, the smallest ster-

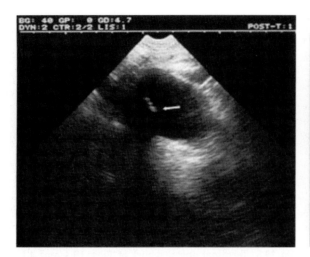

Fig. 19.7. An 80-year-old female with violent chest pain. Suprasternal scan demonstrating an enlarged aortic lumen with an internal image that is irregular, nonartifactual, and mobile indicating intimal flap (*arrow*). Dissection of the thoracic aorta

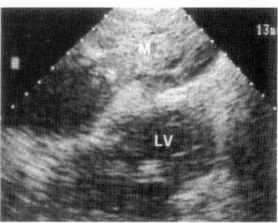

Fig. 19.8. Substantial collection (*M*) visible by the transsternal route, in a recently operated patient. The collection is echoic and tissue-like. The tap withdrew frank pus. Note the heart (*LV*) located more deeply

nal disunity can offer a large route for the ultrasound beam. Acute mediastinitis can then be diagnosed. In a patient who had sepsis 1 month after aortic dissection cure, the transsternal route showed a large, echoic mass of the retrosternal space (Fig. 19.8). An ultrasound-guided puncture of this mass immediately withdrew frank pus. Staphylococcus was isolated in a few minutes by the laboratory, and adapted antibiotic therapy was begun before prompt surgery.

Acute mediastinitis can often be diagnosed by the anterior parasternal route, if the collection is anterior and voluminous, and extends beyond the sternum.

In mediastinitis with the thorax opened, we have not yet seen an advantage to in situ ultrasound analysis. If indicated, the probe can be inserted in a sterile sheath.

It is assumed that the possibility of early diagnosis of acute mediastinitis by transesophageal echography is promising.

Thoracic Esophagus

Thoracic esophagus cannot be explored by a retrotracheal approach. It can be approached below the carina as a tubular flattened structure that passes in the dihedral angle between the heart and descending aorta (Fig. 19.9). Its analysis is uncertain but should always be tried.

Esophageal rupture is an emergency whose infrequency makes it all the more severe, since this diagnosis is rarely evoked immediately. Our observations show that a routine ultrasound examination of any thoracic drama will promptly recognize these disorders: partial pneumothorax, pleural effusion (with alimentary particles yielding a complex echostructure), and frank pus withdrawn from the ultrasound-guided thoracentesis.

In the critically ill patient, the gastric tube and above all its frank acoustic shadow make a good

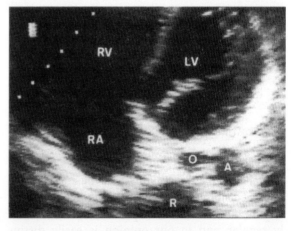

Fig. 19.9. Location of the thoracic esophagus (*O*) in a transverse, pseudo-apical scan of the heart. The esophagus is surrounded by the rachis (*R*), the right auricle (*RA*), the left ventricle (*LV*) and the descending aorta (*A*)

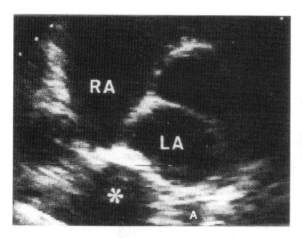

Fig. 19.10. Inflated esophageal balloon of a Blakemore probe (*asterisk*), driving the posterior aspect of the left auricle (*LA*) away

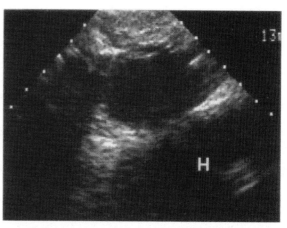

Fig. 19.12. False aneurysm of the left internal mammary artery. Transverse scan of a parasternal intercostal space. Egg-shaped mass with vertical long axis. In real-time, an echoic whirling flow indicated the arterial nature of this mass. *H*, heart

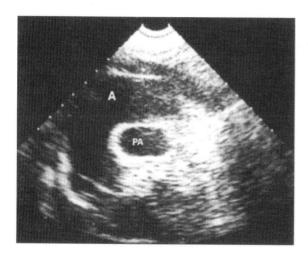

Fig. 19.11. Another transverse scan of the right pulmonary artery (*PA*), surrounded by the aortic arch (*A*). Suprasternal scan. A pulmonary embolism could thus be proven in extreme emergency

landmark that facilitates the location of the esophagus. The esophageal balloon of a Blakemore tube can be visualized posterior to the left auricle (Fig. 19.10). Ultrasound help in this situation is discussed in Chap. 6.

Pulmonary Artery

In patients with a favorable morphotype, the aortic arch can be exposed by suprasternal route. In the concavity of the aorta, a transverse scan can more or less easily bring the right pulmonary artery into view (Fig. 19.11). Detection of a frank blood clot using this route is rare, but can provide immediate diagnosis of severe pulmonary embolism.

Internal Mammary Artery

The internal mammary artery crosses just outside the sternal border. Locating it can be useful before certain punctures.

An internal mammary artery false aneurysm once had this very suggestive pattern: an egg-shaped, vertical, long-axis mass. Ultrasound analysis of its content showed a blatant whirling flow (Fig. 19.12). The vascular origin of this mass was proven, once again without Doppler. It goes without saying that this pattern seriously contraindicates diagnostic puncture.

Other Mediastinal Structures

The recognition of the following elements, even if they are responsible for disorders such as tracheal compression, rarely leads to therapeutic decisions in the emergency room. Diving goiter, adenomegaly or mediastinal tumors can be quietly diagnosed when not compressive (see Fig. 12.8, p 73). An anterior mediastinal mass in a clinical context of myasthenia gravis will be suggestive of thymoma. A pneumomediastinum yields, in our observations, a complete acoustic barrier, of value if (1) the heart was previously located in this area and (2) lung sliding is recognized outside this area, which rules out pneumothorax.

Let us remind the reader here that complete atelectasis can considerably favor the ultrasound analysis of the mediastinum by the external approach (see Fig. 12.20, p 80, and Fig. 17.11, p 124).

References

1. Matter D, Sick H, Koritke JG, Warter P (1987) A suprasternal approach to the mediastinum using real-time ultrasonography, echoanatomic correlations. Eur J Radiol 7:11–17

Chapter 20

General Ultrasound of the Heart

We could have placed the heart first, because of its strategic importance, or last (another mark of recognition). As the heart can be considered a vital ultrasound-accessible organ like others, a logical place was here.

Obviously, reference textbooks treat this subject exhaustively [1, 2]. The notions which follow are intentionally simplified to the maximum in a double aim: to remain faithful to the title of the book (hence the title of this chapter) and, as a consequence, be able to show to a non-cardiologist some of the characteristic features seen in the emergency situation: left heart hypokinesis, right heart dilatation, pericardial tamponade, hypovolemic shock, etc. The physician who is not a cardiologist examines the heart, then requests confirmation from a specialist – unless time does not allow. Because time is always a critical issue, an intensivist trained in emergency ultrasound should clearly be trained in applying the probe on the heart.

The reader is therefore invited to acquire the basic knowledge necessary. This chapter could have been written by a cardiologist. Yet echocardiography is usually done using sophisticated material and highly trained personnel, with complex thought processes. Simple material and a simple technique can yield useful information. Having accrued experience in a pioneering institution in echocardiography since 1989, the authors have come to the tentative conclusion, open to consideration, that simple therapeutic procedures can be deduced from the observation of simple phenomena. For instance, Chap. 28 shows that, in the precise setting of searching for the origin of acute dyspnea, extremely limited investigation of the heart can be sufficient: in particular, the right ventricle status can be deduced from lung analysis.

Deliberating on echocardiography without mentioning the Doppler in 2004 may appear overly bold and thus requires explanation. The drawbacks of Doppler equipment were detailed in Chap. 2. All ICUs are not equipped with transesophageal Doppler echocardiography – far from it – most are even not equipped with simple units. Some new techniques such as the PICCO aim to replace echocardiography. Yet a simple, two-dimensional heart examination will give vital information in the emergency setting.

Let us recall a basic point: acquiring an ultrasound dedicated to the transesophageal route blocks the way to general ultrasound and condemns the user to visualizing only the heart. The reader will therefore not take offense if transesophageal ultrasonography is not discussed in this chapter. Here again, reference textbooks exist on this semi-invasive technique. Even minimal but basic information can always or nearly always be extracted from a surface ultrasound examination [3].

One advantage of Doppler is monitoring cardiac output using an endoesophageal system [4]. The question of whether these parameters are mandatory in emergency care is a source of controversy [5]. Rather than sustaining these controversies, we suggest one basic point: two-dimensional ultrasound cannot give parameters obtained by invasive or semi-invasive techniques. However, it integrates data that are not only cardiac, but also venous, abdominal (inferior vena cava) and above all pulmonary (status of the artifacts). The level of investigation will be altered in such a way that the amount of information lost in hemodynamic terms is regained in terms of diagnosis. For example, fine analysis of the Doppler signal of the pulmonary veins in a critically ill patient can be less useful if one has made the diagnosis of tension pneumothorax, for instance, or massive pulmonary embolism. In other words, our logic is to favor the urgent needs first.

Finally, it must be noted that the hemodynamic investigation, either invasive (Swan-Ganz) or semi-invasive (transesophageal echocardiography) leads to three simple alternatives: whether to give fluid therapy, inotropic agents, or vasopressors. It is of great interest to note that surface examination, including the heart but also the lungs and veins, can be compared with these complex approaches when only the medical prescription changes are taken into account. In terms of therapeutic impact, our daily experience is edifying (study in progress).

Heart Routes

The parasternal route lies in the left parasternal area. The apical route corresponds to systolic shock. Positioning the patient in the left lateral decubitus, the reference in cardiology, is not always easy in a critically ill patient (Fig. 20.1).

Mechanical ventilation often creates a barrier to the transthoracic approach of the heart. Fortunately, the subcostal route is a frequent answer to the poor-quality images resulting from the thoracic routes. This route is widely employed in the intensive care unit in sedated supine patients. This is an abdominal approach, with the probe applied just to the xiphoid, the body of the probe applied almost against the abdomen.

It is rare that cardiac function cannot be assessed in the emergency situation. Several techniques can be used. In the parasternal approach, for instance, care should be taken to wait for the end-expiratory phase. It is often possible to obtain, even if only for a fraction of a second, a dynamic image of the heart that suffices for a rough evaluation of the left heart status. If needed, one can lower the respiratory rate for a short time in order to prolong this instant. The quality of the subcostal route is improved if the hepatic parenchyma is used as an acoustic window. Therefore, in some instances the probe should be moved far from the thorax. A right intercostal approach through the liver can analyze the auricles, or even more. This route (not yet described to our knowledge) should be tried when no other route is possible. The stomach can be filled with fluid in order to create an acoustic window making the subcostal approach easier. A right parasternal approach will be contributive if the right chambers are dilated and extend to the right.

All these techniques, when they provide an answer to the clinical question, should theoretically decrease the need for the transesophageal technique. Above all, they respond to a precise philosophy: simplicity. If this approach has answered the question, one can consider that simplicity was the winning choice.

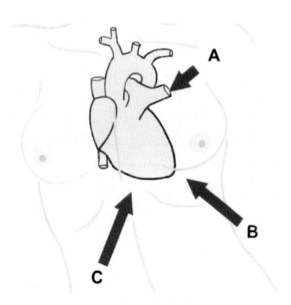

Fig. 20.1. The three classic routes of the heart. *A* The parasternal route. *B* The apical route. *C* The subcostal route, a basic approach to the ventilated patient

Notions of Ultrasound Anatomy of the Heart

The heart is a complex mass, which one can schematize from the left ventricle. The left ventricle is like an egg-shaped mass with a long axis pointing leftward, downward and forward. It has a base (where the aorta and left auricle are located), an apex, and four walls: inferior, lateral, anterior, and the right wall, which is called the septal wall. This wall is made by the septum. The right ventricle has more complex anatomy. Its apex covers the septum, its base (infundibulum) covers the initial aorta. It has a septal wall and a free wall. Intracavitary structures are the valves and the left ventricular pillars. The auricles are visible behind the ventricles. The cardiac muscle is echoic. The chambers are anechoic (except for situations of cardiac arrest).

An excellent way to learn heart anatomy is to use ultrasound, since it reduces a rather complex three-dimensional structure to more simple two-dimensional structures.

Normal Ultrasound Anatomy of the Heart

- The parasternal route, long-axis view, studies the left ventricle (except the apex), the left auricle, the initial aorta, the right ventricular infundibulum, and the dynamics of the mitral and aortic valves (Fig. 20.2).

Normal Ultrasound Anatomy of the Heart   141

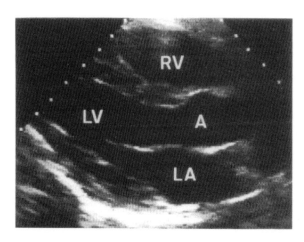

Fig. 20.2. Long-axis view of the heart, left parasternal route. A concession to cardiology was made, since this figure is oriented with the patient's head at the right of the image. *LA*, left auricle; *LV*, left ventricle; *RV*, right ventricle; *A*, initial aorta

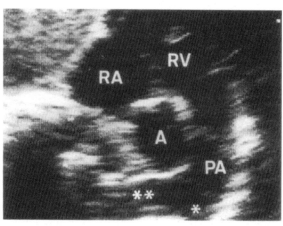

Fig. 20.4. Small-axis parasternal view of the base. *RA*, right auricle; *RV*, right ventricle, prolonging by the pulmonary artery (*PA*), which surrounds the initial aorta (*A*). Right (**) and left (*) branches of the pulmonary artery

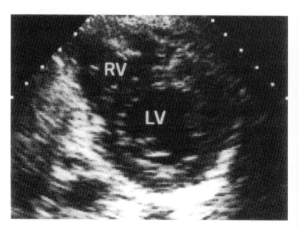

Fig. 20.3. Small-axis biventricular parasternal view. The left ventricle (*LV*) section is round. The two prominent structures are the pillars of the mitral valve. The right ventricle (*RV*) surrounds the septal aspect of the left ventricle

- The parasternal route, short-axis view, studies the two ventricles and the septum at the bottom (Fig. 20.3). Higher up, it shows a view where the right auricle, the tricuspid valve, the basal portion of the right ventricle, the pulmonary artery and its two division branches, which surround the initial aorta, are visible (Fig. 20.4).
- The apical route, four-chamber view, provides an overview of the four chambers. This view gives the most information, and shows the heart in its true symmetry axis: ventricles anterior and auricles posterior, left chambers to the right, right chambers to the left (Fig. 20.5). The

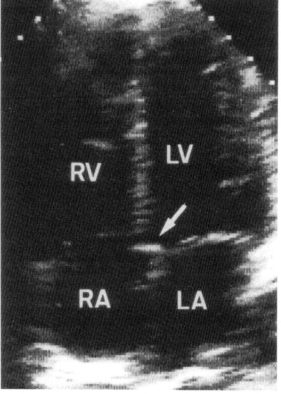

Fig. 20.5. Four-chamber view, apical window. Here, the heart seems to be a symmetric structure. *LV*, left ventricle, *LA*, left auricle, *RV*, right ventricle; *RA*, right auricle. This incidence allows immediate comparison of the volume and dynamics of each chamber. Note that the plane of the tricuspid valve is more anterior than the plane of the mitral valve. In other words, right auricle and left ventricle are in contact (*arrow*), a detail which allows correct recognition of each chamber

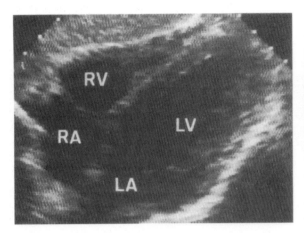

Fig. 20.6. Subcostal view of the heart. This approach is a classic in the intensive care unit. It is a truncated equivalent of the four-chamber apical view in Fig. 20.5. *RV*, right ventricle; *RA*, right auricle; *LV*, left ventricle; *LA*, left auricle. This fixed image is insufficient and the operator must scan this area by pivoting the probe from top to bottom to acquire a correct three-dimensional representation of the volumes. The pericardium is virtual here

lateral and septal walls and the apex of the left ventricle are visible.
- The apical route, two-chamber view, is obtained by rotating the probe 90° on its long axis, and allows analysis of the anterior and inferior walls of the left ventricle.
- The subcostal route gives a truncated view of the heart. It thus cannot help for precise measurements. However, this route is easily accessible in a critically ill patient and thus is of major interest (Fig. 20.6). An overview of the pericardial status, chamber volume and myocardial performance is available.

All the routes allow analysis of the pericardium, normally virtual or quasi-virtual.

Normal Measurements

Static Measurements

Only rough estimates will be given. In a short axis at the pillar level, the left ventricular walls (septal or posterior) are 6–11 mm thick in diastole. The left ventricle chamber caliper is 38–56 mm. The right ventricle free wall is less than 5 mm thick, but a precise measurement should include subtle criteria, since the shape of the right ventricle is complex. In an apical four-chamber view, the right ventricle size is less than that of the left ventricle.

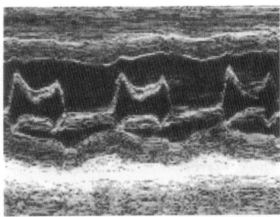

Fig. 20.7. Time-motion recording of the mitral valve. A kind of »M« is displayed inside the left ventricle. Long-axis parasternal view

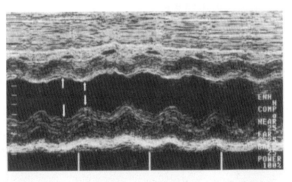

Fig. 20.8. When the left ventricle is bisected by the time-motion line (see Fig. 20.3), its contractility can be objectified on paper. The narrower the sinusoid wave, the more the contractility is decreased. If precise data are preferred to a visual impression, a very rigorous technique is required, using a perfectly perpendicular axis, thus avoiding distortions due to tangency, and a measure between pillars and coaptation of the mitral valve, a reproducible area. The *arrows* indicate diastolic then systolic diameter of the left ventricle. The contractility is normal here, not exaggerated (shortening fraction, 28%). Muscle thickness variations may also be measured on this figure

Dynamic Measurements

Real-time analysis allows appreciation of the ventricular contractility and, more secondarily for us, wall thickening and valve movements (Fig. 20.7). A time-motion image through the ventricular small axis can measure (Fig. 20.8):

- The left ventricular chamber caliper in diastole, which indicates whether there is dilatation.
- This caliper in systole, which defines contractility. The difference of these two values, divided

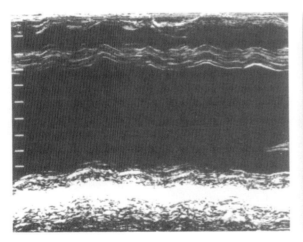

Fig. 20.9. Left ventricle hypocontractility. The sinusoid wave is near the horizontal line in this patient with cardiac failure because of dilated cardiomyopathy (diastolic diameter, 67 mm)

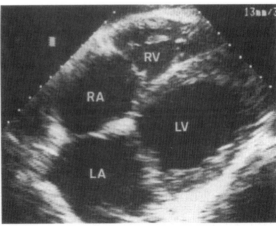

Fig. 20.10. Dilated cardiomyopathy, with massive enlargement of the four chambers

by the diastolic caliper, defines the left ventricle shortening fraction, a basic parameter of the ventricular systolic function. It is normally 28–38%. This information does not replace the ejection fraction, but it is easy to obtain in the emergency situation.

- The parietal thickening fraction (the ratio of the difference of diastolic and systolic thickening over diastolic thickening, normal range from 50% to 100%) is less useful in our daily (and above all nighttime) routine.

The changes in these parameters is assessed with treatment.

Left Ventricular Failure

When systolic function is impaired, global contractility is decreased, with low shortening fraction (Fig. 20.9). This profile can be seen in left ventricular failure of ischemic origin, dilated cardiomyopathies (Fig. 20.10), septic shock with heart failure, and drug poisoning from carbamates with heart injury.

The impairment of the diastolic function of the left ventricle is more delicate to detect if Doppler is not used. However, in a certain percentage of cases, diastolic dysfunction is due to myocardial hypertrophy. This profile, which is accessible to simple two-dimensional ultrasound, can provide a strong argument for this etiology (Fig. 20.11).

It should be stated here that in a patient suspected of pulmonary edema, the usual procedure

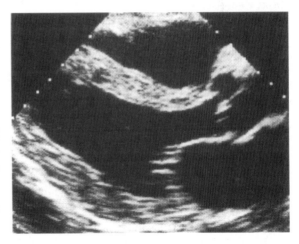

Fig. 20.11. Left ventricle hypertrophy with parietal thickness at 20 mm. A sort of parietal shock was perceived in this patient (not reproduced here since there was no time-motion acquisition). It was synchronized with the auricle systole and probably indicated a sudden increase in pressure in a chamber whose volume could not increase. Long-axis parasternal view

is to search for cardiac failure. However, an initial step would sometimes avoid faulty shuntings: first checking for pulmonary edema by searching for lung rockets (see Chap. 17). An absence of lung rockets means no pulmonary edema. Lung rockets give qualitative information on capillary wedge pressure and may also be useful in measuring lung water (Chap. 17, p 122).

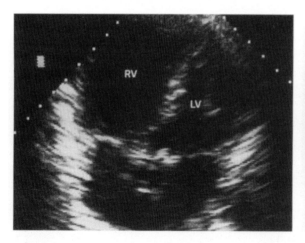

Fig. 20.12. Massive dilatation of the right ventricle in a four-chamber view using the apical route. Massive pulmonary embolism

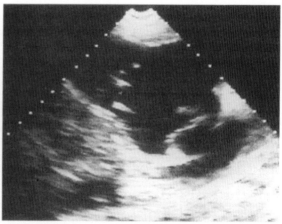

Fig. 20.13. Peculiar pattern evoking a royal python's head. It is in fact a parasternal long-axis view of a massively dilated right ventricle. Young patient with ARDS

Right Heart Failure

In normal conditions, the right ventricle works under a low-pressure system. Any hindrance to right ventricular ejection will quickly generate dilatation [1]. Acute right heart failure associates early right ventricular dilatation, a displacement of the septum to the left, and a tricuspid regurgitation. This regurgitation can, if needed, be objectified without Doppler, in patients with a spontaneously echoic flow: analysis of the inferior vena cava will show this particular dynamics. The free wall of the right ventricle is not thickened in case of a recent obstacle.

This ultrasound pattern can be seen in severe asthma, adult respiratory distress syndrome, extensive pneumonia, and in pulmonary embolism with hemodynamic disorders (Figs. 20.12, 20.13).

If the right heart is not accessible to transthoracic ultrasound, note that numerous diagnoses of acute dyspnea can nonetheless be made (see Chaps. 18 and 28).

Chronic pulmonary diseases generate adaptation of the right heart muscle, and COPD patients with acute exacerbation will also have thickened free wall. The dilatation is often major (Fig. 20.14).

Pulmonary Embolism

The characteristic ultrasound features are described in Chaps. 17, 18 and 28. In our approach, for the diagnosis of pulmonary embolism alone, heart analysis has a small place. Analysis of the lung sur-

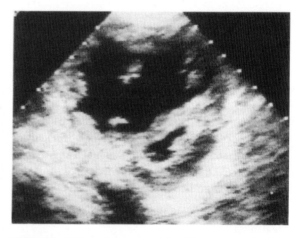

Fig. 20.14. Major right ventricle dilatation with flattening of the left ventricle. Note the substantial thickening of the free wall of the right ventricle. Short-axis parasternal view

face and the venous system (inferior as well as superior) contribute major information. Note that the echocardiographic findings of pulmonary embolism are nonspecific, as they are common to a number of causes of acute right ventricular pressure overload such as the ARDS or status asthmaticus [6]. The lung pattern, if normal in a dyspneic patient, is predictive of right heart failure. The combination of a normal lung pattern with venous thrombosis in a dyspneic patient is highly characteristic, and precious time can be saved. Our experience shows that most patients can be treated in the emergency room before invasive steps are taken. During a transthoracic examination, observation of a blood clot in the right pulmonary artery may appear anecdotal, but when present, the diagnosis of pulmonary embolism should be considered definite (see Fig. 19.11).

The diagnosis of pulmonary embolism in a patient with ARDS may be a challenge – if venous thrombosis is no longer visible [6]. Here, the transesophageal approach should be accorded its proper place. Ultrasound proof of embolism is the visualization of an embolus in a main pulmonary artery [7]. The endovascular ultrasound approach was proposed long ago and may also provide a bedside diagnosis [8].

Acute Pericarditis

This diagnosis is a basic illustration of the concept of general ultrasound of the heart, a diagnosis which should be within every intensivist's reach. The two layers of the pericardium are separated by a more or less anechoic collection. A sort of sinusoid is visible, i.e., the thickness of the effusion varies during the heart cycle. A pericardial effusion is first detected posterior to the left ventricle, then anterior to the right ventricle, then becomes circumferential.

Hemopericardium can be more or less echoic. Purulent pericarditis can contain visible septations (Fig. 20.15).

Pericardial Tamponade

When pericardial effusion is detected in an unstable patient, the possibility of tamponade must be raised.

A pericardial tamponade is always abundant and circumferential (Fig. 20.16) except in some postoperative cases, where small effusions can have consequences.

Within a distended pericardial sac, the heart appears to be swimming.

The description of minute signs using Doppler data will have two effects. One effect will be beneficial: the tamponade feature will be highlighted. One effect will be deleterious: time will be lost in searching for a specialist or opportunities will be lost if the logistics (trained operator, sophisticated unit) are not present on site. As always in emergency situations, the place for academic approaches is limited. The opportunity to insert a needle, monitored by ultrasound, in an unstable patient with abundant pericardial effusion will be less

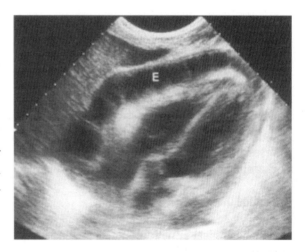

Fig. 20.15. Fluid collection in the pericardial space (*E*). The septations indicate an infectious cause. Note that the effusion surrounds the entire heart: it is visible posterior to the left ventricle in this subcostal approach. Pleuropericarditis due to pneumococcus

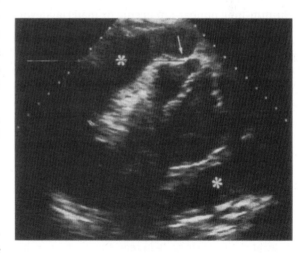

Fig. 20.16. Pericardial tamponade. The heart is surrounded by an abundant fluid collection (∗). A swinging pattern was visible in real-time. The right chambers are collapsed, with collapse of the right ventricle free wall (*arrow*). This subcostal figure also shows the route for a life-saving tap

often missed than any loss of time or intellectual attitude. In other words, in such patients, there is little place for other diagnoses.

Time permitting, simple devices make it possible to observe signs in rhythm with cardiac and respiratory cycles in the spontaneously breathing patient:

- Inspiration facilitates venous return, and the right ventricle dilates at the expense of the septum, which is more compliant than the free wall. The septum is shifted to the left and compresses

the left ventricular chamber (hence the pulsus paradoxus).
- Diastole creates a decrease in intracavitary pressures, whereas intrapericardial pressure remains constant. The right chambers are thus collapsed by the surrounding pressure. The right auricle wall collapses first, an early sign. Then, the free wall of the right ventricle is involved. In extreme cases, right chambers can be undetectable. Right-chamber collapse is amplified by hypovolemia.

Pericardial Drainage

When the clinical situation is critical, ultrasound allows an immediate and safe pericardial tap. A minor fluid withdrawal can dramatically improve the circulatory status.

In such situations, the type of material does not matter. If there is evidence of viscous fluid, a large caliper of needle will be preferred. One can use a 90-mm-long lumbar tap needle or material devoted to thoracentesis such as the Pleurocath, a thin chest tube.. Such materials should at best remain permanently on the trolley, in a dedicated place.

The pericardium is best approached via the subcostal route with ultrasound guidance. The probe is applied next to the needle, which should be inserted in the plane of the probe. An aseptic technique depends on the clinical situation (i.e., minimal in case of cardiac arrest). As for any ultrasound-guided procedure, the progression of the needle can be followed through the liver parenchyma (Fig. 20.17). When the tip of the needle is located in the fluid collection, a second operator aspirates the syringe, whereas the first operator firmly maintains the needle under permanent control on the screen, since the heart is not far. If blood is withdrawn, the second operator reinjects it without disconnecting the syringe. If this blood originated from the pericardial sac, this maneuver creates visible echoic turbulence within the collection. This turbulence cannot be seen if the blood comes from a heart chamber, a situation which should not occur if the ultrasound guidance is effective. Microbubbles (contrast echography) can also be used, time permitting.

Hypovolemic Shock

Comments on the role of general ultrasound in assessing blood volume are available in Chap. 28. When all ultrasound data agree, the typical profile

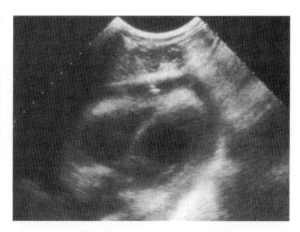

Fig. 20.17. Ultrasound-guided pericardial tap via the subcostal approach. The needle is totally visualized within the hepatic parenchyma when penetrating the pericardial cavity. Purulent pericarditis due to pneumococcus

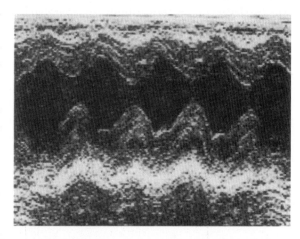

Fig. 20.18. Hypercontractile pattern of the left ventricle during hypovolemic shock. Time-motion acquisition in a short-axis parasternal view. Small diastolic chamber. Quasi-virtual systolic chamber. Tachycardia

includes hypercontractile left ventricle, with small or sometimes virtual end-systolic chamber volume (Fig. 20.18), flattened inferior vena cava, and a lung surface free of any lung rockets, especially in the dependent areas.

Gas Tamponade

With the ultrasound device immediately available, in a critical situation it is possible to immediately detect collapsed chambers, without pericardial effusion. The subcostal approach is usually the only contributive approach. This pattern will

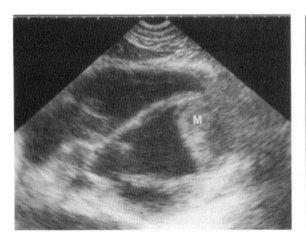

Fig. 20.19. Voluminous thrombosis (*M*) at the left ventricle apex. Subcostal view

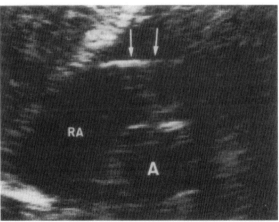

Fig. 20.20. Partial visualization of a Swan-Ganz catheter in the right ventricle. The balloon inflation and the route of the catheter toward the pulmonary artery can be followed on the screen

contrast with enlarged jugular veins found clinically or with ultrasound. Searching for bilateral, compressive pneumothorax completes this exploration.

Intracavitary Devices

Intracardiac thromboses can be identified (Fig. 20.19). They give a regular echoic pattern, sometimes mobile. A foreign body such as the distal end of a catheter should be searched for in the right chambers. These applications are highly dependent on the quality of the available windows. More interesting is the ability to check, in real-time, the proper position of a electrosystolic probe in the right ventricle. If a Swan-Ganz catheter is judged necessary, its on-target progression within the pulmonary artery can be verified with appropriate incidences (Fig. 20.20). We have proceeded with two operators. One, sterile, inserts the material, the other guides the distal end of the catheter using the subcostal approach. Asepsis can be efficiently controlled.

Gas Embolism

The ultrasound diagnosis of gas embolism is possible at the heart (Fig. 20.21). Gas embolism yields large, rough hyperechoic echoes, with posterior shadow, and is highly dynamic. In a supine patient, these gas bubbles collect at the anterior part of the right ventricle and travel little by little in the pulmonary artery – unless the patient is promptly turned to the left lateral decubitus position. Gas embolism complicating the insertion of a venous central catheter can be predicted when an inspiratory venous collapse is identified (see Chap. 12).

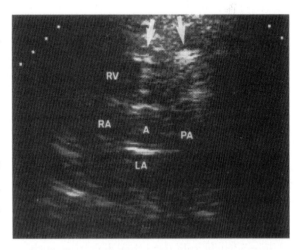

Fig. 20.21. Gas embolism. In this short-axis parasternal view of the base, real-time visualization allows immediate diagnosis. Hyperechoic images (*arrows*) are identified at the roof of the right ventricle (*RV*), highly mobile, as gas bubbles can be within a dynamic hydraulic circuit. They repeatedly appear and progressively are drawn toward the pulmonary artery (*PA*). Suboptimal quality figure, obtained in emergency setting. *RA*, right auricle; *LA*, left auricle; *A*, aorta

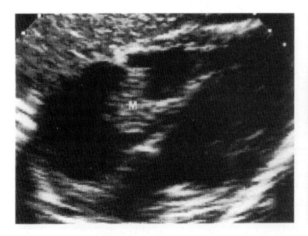
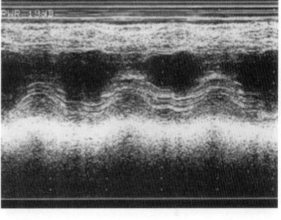

Fig. 20.22. Tissue-like mass depending on the tricuspid valve. A diagnosis of endocarditis in a young drug addict was immediately made using this subcostal ultrasound view, quickly confirmed by positive hemocultures (staphylococcus). *M*, vegetation

Fig. 20.23. Frank akinesia of the septal wall of the left ventricle. Short-axis parasternal view, time-motion mode. Note the hypercontractility of the lateral wall, located opposite the septal wall. Anteroseptal myocardial infarction seen at the 3rd h

Endocarditis

Endocarditis can be suspected when an echoic image, arising at the free part of a valve, can be detected (Fig. 20.22). Traditionally, the gold standard is the transesophageal approach. However, in our experience, the cases we have encountered all gave signs that were already very suggestive, not to say specific, before being confirmed by semi-invasive or invasive procedures.

Myocardial Infarction

The ischemic wall is motionless, which contrasts with the normal or exaggerated dynamics of the other walls. This pattern is not always characteristic. The diagnosis of segmental anomalies is often subtle and certainly requires extensive experience (Fig. 20.23).

The investment is worthwhile if it is accepted that ultrasound anomalies are visible very early, thus altering immediate management [9]. Emergency ultrasound in a patient with suspected myocardial infarction has the merit of being able to immediately rule out other diagnoses such as pericarditis or sometimes aortic dissection, whose management differs.

Cardiac Arrest

Using emergency ultrasound to detect cardiac arrest should become routine in the years to come (Fig. 20.24). The wide-ranging potentials of a simple device are detailed in Chap. 28.

Miscellaneous

Many anecdotal situations can be encountered in the emergency room, but their exhaustive description would overburden this book. To give one example, in a young woman admitted for severe shock with massive pulmonary edema, transthoracic ultrasound objectified a retroauricular mass

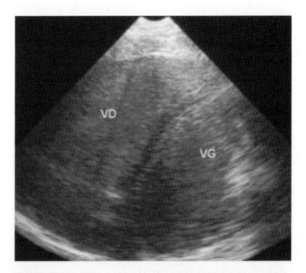

Fig. 20.24. In this subcostal view, all chambers have echoic homogeneous content. This sludge pattern is the result of cardiac arrest. The chambers will become normally anechoic after recovery of a cardiac activity

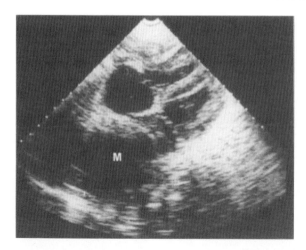

Fig. 20.25. Subcostal view of a young woman in shock with white lungs. A mass is visible at the location of the left auricle (*M*). This is an esophageal abscess that complicated local surgery performed weeks before. The shock was caused by septic disorders (with positive hemocultures) as well as by a hindrance to pulmonary venous return. We reassure the reader: the diagnosis was not immediate but rather perioperative (an emergency transesophageal ultrasound examination was also performed, and was ineffective as well)

compressing the left auricle (Fig. 20.25). Emergency surgery revealed an esophageal abscess, which was responsible for both septic shock and hemodynamic failure due to impairment in pulmonary venous return.

Valvular diseases, problems with mechanical valves, certain mechanical complications of myocardial infarct, hypertrophic asymmetric cardiomyopathies cannot be described here. Numerous subtleties depending on Doppler would also be beyond the scope of this book. Specialized techniques such as transesophageal Doppler echocardiography, used by specialists, will provide the best logistical conditions [10].

In Conclusion

Let us recall that the device described in Chap. 2 is appropriate for two-dimensional cardiac imaging.

Moreover, the approach described here is simplified. General ultrasound of the heart does not provide the same information as transesophageal echocardiography, but it does not answer the same questions, and is performed for different purposes. Finally, integrating this simplified cardiac approach into a whole-body framework, including in particular the lung and venous status, will provide basic information. In the emergency situation, this information allows an investigation at a level close to, and in certain cases better than, the traditional approach, which is based on the heart alone and can sometimes suffer from inadequacy. Studies in progress will soon confirm this belief.

References

1. Jardin F, Dubourg O (1986) L'exploration échocardiographique en médecine d'urgence. Masson, Paris
2. Braunwald E (1992) Heart disease. Saunders, Philadelphia
3. Vignon P, Mentec H, Terré S, Gastinne H, Guéret P, Lemaire F (1994) Diagnostic accuracy and therapeutic impact of transthoracic and transesophageal echocardiography in mechanically ventilated patients in the ICU. Chest 106:1829-1834
4. Diebold B (1990) Intérêt de l'échocardiographie Doppler en réanimation. Réan Soins Int Med Urg 6:501-507
5. Jardin F (1997) PEEP, tricuspid regurgitation and cardiac output. Intensive Care Med 23:806-807
6. Schmidt GA (1998) Pulmonary embolic disorders. In: Hall JB, Schmidt GA, Wood LDH (eds) Principles of critical care, 2nd edn. McGraw Hill, New York, pp 427-449
7. Goldhaber SZ (2002) Echocardiography in the management of pulmonary embolism. Ann Intern Med 136:691-700
8. Tapson VF, Davidson CJ, Kisslo KB, Stack RS (1994) Rapid visualization of massive pulmonary emboli utilizing intravascular ultrasound. Chest 105:888-890
9. Horowitz RS, Morganroth J, Parrotto C, Chen CC, Soffer J, Pauletto FJ (1982) Immediate diagnosis of acute myocardial infarction by two-dimensional echocardiography. Circulation 65:323
10. Vignon P, Goarin JP (2002) Echocardiographie-Doppler en réanimation, anesthésie et médecine d'urgence. Elsevier, Amsterdam

Chapter 21

Head and Neck

Here again, analysis of a field that is not yet routine in emergency ultrasound can perform unexpected services in the ICU.

Maxillary Sinuses

Maxillary sinusitis is a basic concern in the ventilated patient. It is assumed to give infectious pneumonia [1] and is subject to diagnostic problems: radiographs with a vertical beam cannot detect air–fluid levels, whereas radiographs with a horizontal beam are not yet routine (and remain irradiating). The usual solution is, once again, referring the patient to CT.

If ultrasound can play even a minimal role, this role should be carefully considered. Available data in the literature regard studies conducted in otorhinolaryngological patients with the A mode. We intentionally did not speak of the A mode in Chap. 1, since this technique is extremely abstract if compared with real-time. The opinion was divided on use of the A mode between the advocates [2, 3] and those preferring a cautious outlook [4]. We have previously written, in error, that conclusion on this aspect was impossible, until the day when, applying our probe on the paranasal areas, we were surprised to see an anatomical view of a maxillary sinus on the screen. This proved that the ultrasound beams were able to cross bones. Note that the scapula or the iliac aisle also do not hinder the beam.

The probe is transversally applied on the square area located between the eye, nose and teeth. The normal image is an absence of signal (Fig. 21.1). This is an artifactual image that is not a posterior shadow, as a bone would generate, but a repetition echo, with dark and clear lines: it is indeed an air artifact. This simple distinction proves that the beam is not stopped by the bone. A pathological signal is the visualization of the sinus itself, i.e., an anechoic image surrounded by two lateral

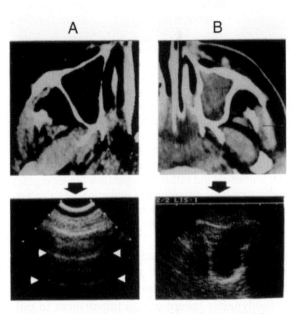

Fig. 21.1. A Normal maxillary sinus. The ultrasound pattern (*top*) is made up of repetition artifacts (*arrows*), which indicate an air barrier. B Total opacity of the sinus. On ultrasound (*top*), the shape of the sinus is outlined: sinusogram in transverse scan. Note the frank pattern, with indicates the total opacity as seen on the CT scan (*top*)

walls and a posterior wall. This pattern was labeled the sinusogram, a self-explanatory term (Fig. 21.1).

Maxillary sinusitis gives a two-step sign. At the first level, there can be a sinusogram, according to an all-or-nothing rule. At the second level, the sinusogram is either complete, with frank visualization of the three walls over the entire area of projection (Fig. 21.1), or incomplete (Fig. 21.2).

One hundred maxillary sinuses of critically ill patients were analyzed in our institution. For simple and clinically relevant correlations, it was necessary to use complex routes, as four pairs of hypotheses were opposed. The relevance of ultrasound is a function of the precision of the words used.

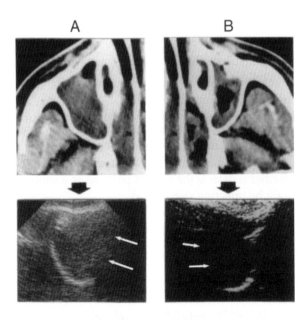

Fig. 21.2A, B. Examples of incomplete sinusograms. **A** This image corresponds to subtotal opacity with a bubble trapped at the top. **B** This one is caused by substantial mucosal thickening. The *white arrows* designate the missing walls, not visualized by the ultrasound

In a ventilated patient, a sinus can (1) be normal, (2) have mucosal thickening, (3) have an air–fluid level, (4) be totally opaque. Only the air–fluid level and total opacity need specific treatment since there is production of pus, logically stemming from drainage.

»Pathological sinus« was an ultrasound term created to designate either hypertrophy on CT, the air–fluid level on CT or total opacity on CT. This term is opposed to »normal sinus«.

»Radiological maxillary sinusitis« is a CT term implying fluid accumulation, i.e., the air–fluid level or total opacity of the sinus. This term contrasts with normal sinus on CT as well as mucosal thickening on CT.

»Total opacity of the sinus« was a CT term implying a complete fluid accumulation. This term contrasts with »normal sinus« and »mucosal thickening«, but also with »sinusitis with the air–fluid level«, a distinction necessary for precise data.

A dynamic maneuver means that the head is in the supine position first, then raised in an upright position. The specification that no dynamic maneuver be done meant that that the patients were studied head supine, as opposed to a dynamic maneuver positioning the head upright.

From these precise definitions, the 100 sinuses comprised 33 radiological maxillary sinusitis cases (with 21 cases of complete opacity), 14 cases of mucosal thickening and 52 normal sinuses. All were studied by CT. Ultrasound performance was as follows [5]:

1. A sinusogram diagnoses pathological maxillary sinus, dynamic maneuvers not taken into account, with a 66% sensitivity and a 100% specificity.
2. A sinusogram diagnoses radiological maxillary sinusitis (vs hypertrophy or normal sinus) with a 67% sensitivity and an 87% specificity, dynamic maneuvers not taken into account.
3. A sinusogram diagnoses total opacity of the sinus, when compared to a partially opacified or mucosal thickening or normal sinus, with a 100% sensitivity and an 86% specificity, dynamic maneuvers not taken into account.
4. A complete sinusogram (as opposed to an incomplete or absent sinusogram) diagnoses total opacity of the sinus (if opposed to partial opacity, i.e., the air–fluid level, hypertrophy or normal sinus) with a 100% sensitivity and a 100% specificity, dynamic maneuvers not taken into account.

In practice, as shown in Table 21.1, a complete sinusogram is specific to total opacity. An incomplete sinusogram, or one detected in a limited area, can indicate either subtotal opacity, with small bubbles trapped against the anterior wall, or substantial mucosal thickening. In a supine patient, the absence of signal can indicate either a normal

Table 21.1. Ultrasound diagnosis of maxillary sinusitis

	Normal sinus	Mucosal thickening	Miscellaneous (polyp)	Maxillary sinusitis (fluid level)	Maxillary sinusitis (total opacity)
Complete sinusogram	0	0	0	0	10
Incomplete sinusogram	0	8	1	2	11
No sinusogram	52	6	0	10	0

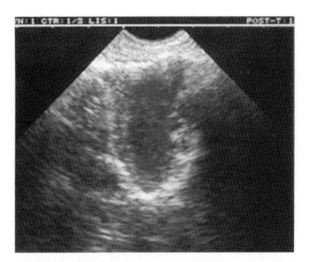

Fig. 21.3. Complete sinusogram. In this case of purulent sinusitis, a double pattern is visible: an internal anechoic area, an external hypoechoic regular frame 4 mm thick. This is an association of a mucosal thickening and a fluid accumulation

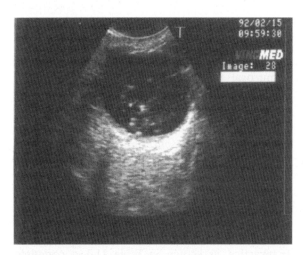

Fig. 21.4. Ophthalmic ultrasound. Multiple echoes as in weightlessness in aqueous humor, which are mobile with the eyeball movements. Vitreal hemorrhage. Diasonic Vingmed unit with a 7.5-MHz probe

sinus or an air–fluid level, which, although substantial, will not be detected if it does not touch the anterior wall.

The diagnosis of air–fluid level (study in progress) requires a dynamic maneuver and gives these signs:

1. No signal with head supine.
2. Detection of a sinusogram in the lower part of the sinus area after positioning the head upright. A small delay will be needed since the fluid can be viscous.
3. Return of the first pattern after positioning the head supine again.

Lastly, subtleties exist in the signs, such as the possibility to differentiate tissue-like hypertrophy from fluid-like sinusitis (Fig. 21.3), a result CT rarely achieves. On the other hand, ultrasound, as well as CT, will not be able to predict the nature of the fluid (pus or blood or noninfected fluid).

Ultrasound is being investigated to determine whether it detects the correct position of a sinusal drain by injecting sterile fluid.

Ultrasound beams cross air (see Chaps. 15–18), they also happen to cross bones.

The Eyeball

The eyeball is accessible through the eyelid, provided no pressure is exerted on the eye so that any vagal reaction is avoided. As for any other examination, the probe is firmly held like a pen, the operator's hand lies firmly on the patient's face, and the probe is gently applied toward the eyelid. The progression of the probe ceases from the instant an image is obtained at the screen. Ophthalmological occult emergencies in comatose patients can therefore be diagnosed (Fig. 21.4).

In case of ocular trauma, a critical issue is the presence of an eye injury. Ultrasound can be contributive, a normal state showing an anechoic, perfectly round organ. In terms of spatial resolution, ultrasound is clearly superior to CT, which irradiates the crystalline lens.

Daily concerns such as the search for ocular candidosis or other disorders may be solved using this technique, we await enough cases to conclude. A retinal hemorrhage would give isoechoic or hyperechoic images anterior to the retina [6].

Optic Nerve and Intracranial Hypertension

The search for intracranial hypertension should ideally be routine in a comatose patient, although it would be inconceivable to perform CT in all comatose patients. However, a system allowing the intensivist to avoid unnecessary erroneous orientation in cases of so-called alcoholic comas, or those that are assumed to be such, would be welcome. The principle of using optic fundus examination was based on the fact that cerebral edema had a centrifuge extension along the optic nerve to the papilla, a clinically accessible area. Meanwhile, CT has replaced this antique examination, which was not sensitive enough. However, this means, once again, transportation of a critically ill patient.

Like any macroscopic structure that is not surrounded by air or bone, the optic nerve is accessible to ultrasound. Yet the optic nerve is an evagination of the brain and is therefore surrounded by meninges. This space is normally virtual. It is logical that any increase in intracerebral pressure will distribute cerebrospinal fluid in all the possible centrifuge directions, including the optic nerve meningeal spaces, even a minute amount. The apparent caliper of the optic nerve will thus be increased.

The technique is detailed in the previous section. The 5-MHz probe can detect, posterior to the eyeball, a sinuous hypoechoic tubular structure that is usually well outlined by hyperechoic fat (Fig. 21.5). Detecting the optic nerve can require some skill. The curves of the nerve must be recognized before any measurement can be taken. If not, in some instances posterior shadows (which are straight) will be confused with the optic nerve. The caliper of the optic nerve can be measured. This caliper is 2.6 mm on average (range, 2.2–3.0 mm). Note that a cardiac probe is totally inappropriate for this application, which requires submillimeter precision.

Before analyzing the data, let us survey some of the theoretical advantages of ultrasound:

1. Bedside technique, immediately implemented.
2. Ultrasound provides in-depth visualization of the optic nerve, whereas optic fundus examination can only analyze the very end of the nerve. Let us imagine, for instance, assessment of the nose. Measuring the length of the nose, should full-face or profile photographs be used? How far does this superiority of ultrasound over optic fundus bring ultrasound compared to CT in the search for intracranial hypertension?
3. Compared with optic fundus examination, ultrasound does not require atropine administration (a time-consuming and sometimes harmful procedure) and is not hindered by cataract.
4. Extremely simple technique. But does it really work?

Let us now analyze our data. On-site observations confirm all these theoretical points. An enlarged optic nerve is pathological (Fig. 21.6). We compared 25 cases of intracranial hypertension proven on CT with 100 critically ill patients with proven

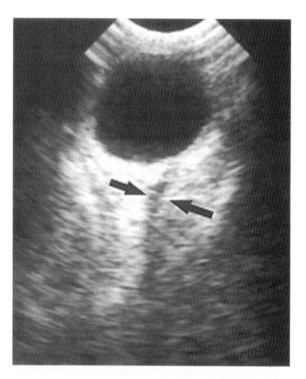

Fig. 21.5. Normal pattern of the eyeball and of the optic nerve in a scan through the eyelid. The optic nerve (*arrows*) has a normal caliper (2.6 mm). Note the sinuous route of the nerve. It should be remembered that the pressure of the probe should be almost null in this kind of approach

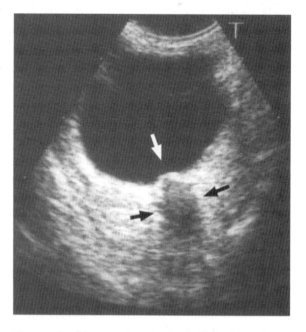

Fig. 21.6. In this scan, the apparent caliper of the optic nerve is markedly enlarged: 5.3 mm (*black arrows*). In addition, the papilla (*white arrow*) bulges in the lumen of the eyeball. There was diffuse brain edema on CT. Diasonic Vingmed unit with a 7.5-MHz probe

absence of intracranial hypertension. Patients with cerebral edema had enlargement of the optic nerve. In this study, the best cut-off was 4.5 mm. Patients who had a greater value had cerebral edema in 80% of cases, patients with a lower value had normal brain status in 83% of cases [7].

Since we prefer values near 100% (see lung ultrasound performance, for instance, in Chaps. 15–18), we were not fully satisfied by these results. Yet several other signs can be analyzed in this field: does the papilla protrude in the eyeball? Is the end of the optic nerve enlarged, bulging or conversely thinned? Is there a visible splitting of the optic nerve? Are the measurements strictly stable or is there imprecision when several measurements are taken? Is there a frank asymmetry between the left and the right? One of these items, or other not yet noted items, may increase ultrasound accuracy: rendezvous in the next edition.

Other applications are being investigated. For instance, we would like to be able to do a spinal tap without losing time in order to check whether this procedure is dangerous. However, it is possible that meningitis always has a minimal degree of intracranial hypertension. This may result in an overly sensitive test. If ultrasound detects minimal brain edema too easily, the benefit of ultrasound may be lost in this particular application. Experience and more extensive data will allow us to conclude.

In practice, when we receive a comatose or encephalopathic patient, we systematically measure the optic nerve. In the absence of strong clinical evidence (of either extreme surgical emergency or ordinary drug poisoning), patients having values below 4.5 mm are monitored at the bedside, and patients with a higher value are referred for emergency CT. Using this policy, we sometimes undertake CT for nothing, but the misdiagnosis of patients with alcoholic coma who do not wake up because they had violent head trauma accompanied by alcoholic intoxication is becoming extremely rare.

The Brain

A probe applied at a precise location of the temporal bone displays a characteristic image, which ends in a structure interpreted as the contralateral bone (Fig. 21.7). This again proves that ultrasound crosses the bones. Brain images, not yet fully identified in the present state of our knowledge, can be described. One aim is to determine whether ultra-

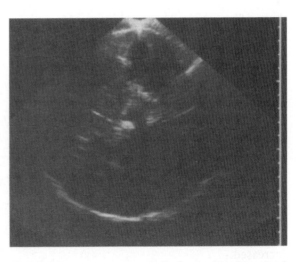

Fig. 21.7. Transverse scan of the brain. The biparietal diameter is 13.5 cm, the usual value in the adult. Many details are visible between the two parietal bones

sound can detect a shift in these images. Earlier, the A mode, a rudimentary ultrasonic system, was used to determine whether the median structures were shifted, thus indicating surgical emergencies [8]. CT now provides accurate answers, but we would be interested to see whether relevant information can be obtained at the bedside, in order to decrease the need for CT in certain instances, or accelerate referral to CT in others.

Data on transcranial Doppler is not included here. This technique is probably of interest in the traumatized patient [9]. We deliberately have not used the Doppler throughout this book, because we believe in a light, unsophisticated, simple and noninvasive tool. In the precise domain of cranial trauma, we maybe commit an injustice. This is why we hope that the measurement of the optic nerve caliper will fill this gap.

The Face

The submaxillary glands and the lingual muscle are accessible using ultrasound. Parotiditis, a classic complication of mechanical ventilation, should give an enlarged, hypoechoic gland, which should be sought between the ear and the maxilla. We lack observations on this obviously rare or perhaps misdiagnosed complication.

The Neck

The neck veins were studied in Chap. 12.

Carotid artery exploration can be useful in a comatose patient. A traumatic dissection will be sought, although the Doppler is the usual technique. Does the two-dimensional approach not give already basic information in some or a majority of cases? This could make the Doppler information redundant in first-line analysis in these cases. Another application of a two-dimensional scanning can be the evaluation of vascular injury by screening for calcifications at the carotid arteries, a marker of the arterial system.

A retropharyngeal abscess can be sought [10]. In this area, traumatic hematomas, other abscesses or cervicofacial cellulitis can be documented. However, CT is preferred here.

The trachea is perfectly detectable at the cervical level: anterior and median with posterior air artifacts. Applying pressure that is more than very light can be very unpleasant in moderately sedated patients. The trachea is quickly lost since it takes a posterior direction when entering the thorax. Via the anterior or lateral approach, one can study its external configuration (Fig. 21.8). Its anteroposterior and lateral diameters can be measured, at inspiration and expiration. Tracheomalacia may be detected this way. Since the tracheal wall is fibrocartilaginous, nothing prevents the ultrasound analysis of the tracheal content: the ultrasound beam encounters the wall, then the air, which stops beam progression. If the anterior wall is thickened by a granuloma or other causes of tracheal stenosis or obstruction, this obstacle will be accurately detected and analyzed. Within the lumen itself, secretions accumulated above an inflated balloon can be detected (Fig. 21.9). This finding may have clinical outcome. Of course, fibroscopy will remain the reference test for tracheal disorders, but the principle remains the same: give the patient a first noninvasive, rapid approach that can alter the usual management, depending on the operator's skill. Some authors use ultrasound for the guidance of percutaneous tracheostomy [11]. The intubation tube itself will give a particular signal, whose clinical application is under investigation.

The thyroid, especially the isthmus, can be usefully located before tracheostomy (Fig. 21.8). An aberrant brachiocephalic artery can be located [12], but also the closeness of the innominate vein or thyroid hypertrophy. Diagnostic ultrasound is

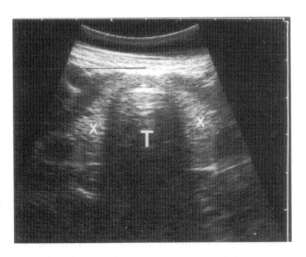

Fig. 21.8. Transverse anterior cervical scan at the thyroid isthmus. The two thyroid lobes (*X*) and the posterior shadow of the trachea (*T*) are recognized. Since an air barrier is visible immediately posterior to the anterior wall of the trachea, it can be possible to conclude that the anterior wall, at this level, is thin

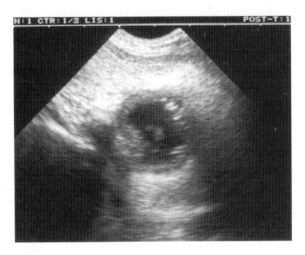

Fig. 21.9. As opposed to Fig. 21.8, this trachea is entirely crossed by the ultrasound beam. There is accumulation of secretions above the inflated balloon. This pattern vanishes if the balloon is deflated, but the patient coughs. In addition, the anterior tracheal wall can be accurately measured, here thickened to 4 mm

contributive if an abnormal thyroid gland is described in a patient with suspicion of severe dysthyroidism. In a young female admitted for acute hypercalcemia, ultrasound immediately detected a suspect mass evoking a parathyroid tumor. This resulted in prompt surgery, which confirmed the diagnosis.

Finally, the rough integrity of the cervical vertebrae can be assessed via the anterolateral cervical

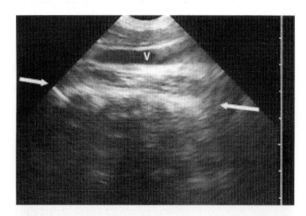

Fig. 21.10. Longitudinal paramedian scan of the neck. Posterior to the internal jugular vein (*V*) and the muscle, a thick hyperechoic line represents the anterior wall of the cervical rachis, here straight without solution of continuity (*arrows*). Upright cervical rachis approach (Fig. 21.10). Why not use first-line ultrasound when there is suspicion of cervical rachis fracture?

For these new fields, of immediate interest in the ICU, high-frequency probes (7.5 or 10 MHz) may be relevant.

The Nape of the Neck

Suboccipital puncture is sometimes performed in patients with intracranial hypertension. Would ultrasound guidance or location be useful in this reputedly risky technique? We are currently investigating the possibilities in this area.

References

1. Rouby JJ, Laurent P, Gosnach M, Cambau E, Lamas G, Zouaoui A, Leguillou JL, Bodin L, Khac TD, Marsault C, Poète P, Nicolas MH, Jarlier V, Viars P (1994) Risk factors and clinical relevance of nosocomial maxillary sinusitis in the critically ill. Am J Respir Crit Care Med 150:776–783
2. Landmann MD (1986) Ultrasound screening for sinus disease. Otolaryngol Head Neck Surg 94:157–161
3. Beuzelin C, Mousset C, Frœhlich P, Senac J, Gory C, Goursot G, Fombeur JP (1990) Evaluation de l'échographie sinusienne dans le diagnostic des sinusites maxillaires purulentes en réanimation. Réan Soins Intens Méd Urg 6:538
4. Rippe JM, Irwin RS, Alpert JS, Fink MP (1991) Intensive care medicine. Little Brown, Boston, p 709
5. Lichtenstein D, Biderman P, Mezière G, Gepner A (1998) The sinusogram: a real-time ultrasound sign of maxillary sinusitis. Intensive Care Med 24:1057–1061
6. Berges O, Torrent M (1986) Echographie de l'œil et de l'orbite. Vigot, Paris
7. Lichtenstein D Bendersky N, Mezière G, Goldstein I (2002) Ultrasound diagnosis of cranial hypertension by measuring optic nerve caliper. Reanimation 11 [Suppl 3]:170
8. Hamburger J (1977) Petite encyclopédie médicale. Flammarion, Paris, pp 1377–1378
9. Czosnyka M, Matta BF, Smielewski P, Kirkpatrick PJ, Pickard JD (1998) Cerebral perfusion pressure in head-injured patients: a noninvasive assessment using transcranial Doppler ultrasonography. J Neurosurg. 88:802–808
10. Rippe JM, Irwin RS, Alpert JS, Fink MP (1991) Intensive care medicine. Little Brown, Boston, p 704
11. Sustic A, Kovac D, Zgaljardic Z, Zupan Z, Krstulovic B (2000) Ultrasound-guided percutaneous dilatational tracheostomy: a safe method to avoid cranial misplacement of the tracheostomy tube. Intensive Care Med 26:1379–1381
12. Hatfield A, Bodenham A (1999) Portable ultrasound of the anterior neck prior to percutaneous dilatational tracheostomy. Anesthesia 54:660–663

Soft Tissues

Soft tissues are accessible to ultrasound. They can be of interest in several instances.

Soft Tissue Abscess

The ultrasound signs include hypoechoic, heterogeneous mass and inconstant punctiform hyperechoic areas indicating bacterial gas (Fig. 22.1), signs indicating a fluid nature such as posterior enhancement (which is inconstant) or changes in dimensions under probe pressure (but such maneuvers can be very harmful, not to say risky). In fact, abscess and hematoma often have similar patterns, and the ultrasound-guided tap will make a definite diagnosis.

Necrotizing Cellulitis

The role that ultrasound can play is not well known in necrotizing cellulitis. The diagnosis is usually clinical. Surgical exploration alone specifies the extension of the necrosis [1]. Ultrasound may theoretically allow early diagnosis by showing deep areas of emphysema before they become clinically accessible. Ultrasound may also distinguish between gangrenous cellulitis (which preserves the muscle) and necrotizing fasciitis (with myonecrosis). Hypoechoic areas dissociating the muscle fibers would then be observed.

Deep Hematoma

A hematoma gives well-limited mass that is anechoic at the first stage and can quickly become echoic and heterogeneous (Fig. 22.2). In case of doubt, ultrasound-guided investigation can give the diagnosis.

A hematoma can develop anywhere and give distinctive signs. At the rectus abdominis muscle, its extraperitoneal nature will be recognized since the peritoneal sliding will be preserved, posterior to the mass. In severe forms, it can be the source of compression (bowel, bladder, etc.) [2].

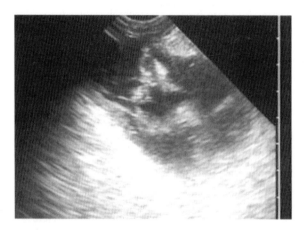

Fig. 22.1. Huge heterogeneous collection in the gluteal area. With ultrasound guidance, the tap withdrew pus, thus confirming the abscess. Young patient with trauma

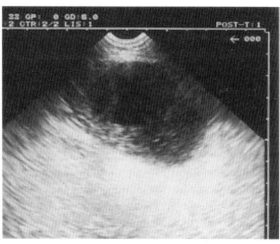

Fig. 22.2. Thigh collection in another traumatized patient. The pattern is not far from that described in Fig. 22.1 but here is a partially solid hematoma

Parietal Emphysema

Parietal emphysema generates air comet-tail-type artifacts. They usually conceal the deeper structures (Fig. 22.3). The presence of parietal emphysema is certainly one of the rare indications to cancel ultrasound examination. However, it is sometimes possible to hide the masses of gas by gentle pressure. At the thoracic level, this is facilitated by the ribs, which remain solid under pressure. Lung sliding can then sometimes be analyzed (see Chap. 16). Note that pneumothorax is not always present.

Let us recall that comet-tail artifacts generated by parietal emphysema can be a dangerous pitfall for the beginner when they appear as E lines. This pattern may be erroneously interpreted as B lines or lung rockets, and genuine pneumothorax can be missed (see Fig. 16.11, p 113). The search for the bat sign in this setting prevents this pitfall.

Edematous Syndromes

In cases of major hydric retention, the soft tissues are enlarged by edema, with hypoechoic zones dissociating the muscles. The analysis of the deeper structures is not hindered, as water is a good conductor for ultrasound beams.

In situations such as nephrotic syndrome with massive hypoalbuminemia, more or less substantial effusions can affect all of the anatomical compartments.

Parietal Vessels

Ultrasound can be useful to accurately locate the epigastric or internal mammary vessels if a local tap is considered (see Fig. 5.12, p 32).

Undernutrition

The nutritional status of a patient is usually monitored by weighing the patient. This is a simple parameter. However, the maneuver is demanding for the paramedical team, and above all, the data obtained is a rough result of inverse trends: in a critically ill patient, the muscles and fat compartments decrease whereas the water compartment increases. Once more, ultrasound can potentially

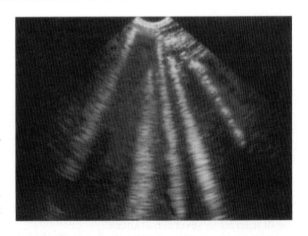

Fig. 22.3. Parietal emphysema. The deep structures in this thoracic view are unrecognizable since they are hidden by numerous comet-tail artifacts. This aspect is unusable. These are W lines, defined as comet-tail artifacts arising from different levels in the soft tissues

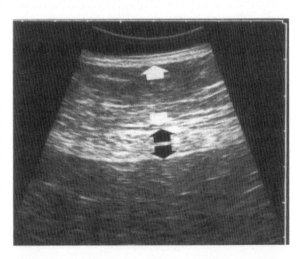

Fig. 22.4. Transverse scan of the paraumbilical abdominal wall. The *white arrows* sharply delimit the fat compartment (17 mm), the *black arrows* the muscular compartment (9 mm for the muscle). These measures can easily be repeated during the stay of the patient. Probe with 7.5-MHz frequency

provide logic-based assistance. A differential analysis of the fat [3], muscle and interstitial compartments can in fact be carried out (Fig. 22.4). Accepting that these variations are the same in any part of the body, only one standardized area should be investigated. An easy-to-access and reliable area is, for instance, a transverse, paraumbilical scan of the rectus abdominis muscle (Fig. 22.4) or, perhaps better, a transverse scan of the crural muscle at mid-thigh. Ultrasound may also detect interstitial edema before clinical evidence, but this precise issue has not yet been investigated.

Miscellaneous

Multiple disorders such as cysts, arterial aneurysms, osteomas, etc. not related to the acute illness can be detected in the soft tissues.

Traumatic Rhabdomyolysis

The muscular loges have increased volume, without abscess or hematoma to explain the clinical swelling. A hypoechoic pattern of the muscles with disorganization of the normal muscular architecture has been described [4]. Another advantage of ultrasound is ruling out associated venous thrombosis (with here a possible place for Doppler if the compression maneuver is harmful).

Malignant Hyperthermia

A heterogeneous and grainy pattern of the muscles, with a hypoechoic pattern of the septa and fascia is described by some [5], not found by others [6]. The rarity of this syndrome in our ICU has until now prevented us from forming an opinion.

References

1. Offenstadt G (1991) Infections des parties molles par les germes anaérobies. Rev Prat 13:1211-1214
2. Blum A, Bui P, Boccaccini H, Bresler L, Claudon M, Boissel P, Regent D (1995) Imagerie des formes graves de l'hématome des grands droits sous anticoagulants. J Radiol 76:267-273
3. Armellini F, Zamboni M, Rigo L, Todesco T, Bergamo-Andreis IA, Procacci C, Bosello O (1990) The contribution of sonography to the measurement of intra-abdominal fat. J Clin Ultrasound 18:563-567
4. Lamminen AE, Hekali PE, Tiula E, Suramo I, Korhola OA (1989) Acute rhabdomyolysis: evaluation with magnetic resonance imaging compared with CT and ultrasonography. Br J Radiol 62:326-331
5. Von Rohden L, Steinbicker V, Krebs P, Wiemann D, Kœditz H (1990) The value of ultrasound for the diagnosis of malignant hyperthermia. J Ultrasound Med 9:291-295
6. Antognini JF, Anderson M, Cronan M, McGahan JP, Gronert GA (1994) Ultrasonography: not useful in detecting susceptibility to malignant hyperthermia. J Ultrasound Med 13:371-374

Part III
Clinical Applications of Ultrasound

CHAPTER 23

Ultrasound in the Surgical Intensive Care Unit

An »echological« distinction between medical and surgical patients should not make sense per se, but some differences can be underlined.

General Issues

The surgical patient is often surrounded by a barrage of acoustic barriers: wounds, dressings, orthopedic material, cervical collar. This may limit the use of ultrasound, but these obstacles can be overcome. The problems of asepsis are more important than in the medical setting, and vigilance regarding crossed infections must be reinforced.

The Abdomen

Dressings sometimes cover the entire abdominal wall, but these limitations can be bypassed. The dressings can be withdrawn, the probe can be inserted in sterile conditions, a sterile contact product can be used, although these procedures may seem overly restrictive. The sterile protection of the probe should conduct the ultrasound beam without interference [1]. Fine transparent adhesive dressings such as OpSite and Tegaderm offer the advantage of being transparent to ultrasound. Their use should therefore be encouraged. Some thick dressings may appear impenetrable by ultrasound, but we have noted that ultrasound beams occasionally are not stopped, and basic answers to clinical questions can be obtained. In addition, medical personnel should be taught to wisely apply dressings, since critically ill postoperative patients will unavoidably have ultrasound examinations.

Apart from the anomalies described in earlier chapters, ultrasound can search for infected postoperative collections [2] (Fig. 23.1). For some authors, ultrasound sensitivity is high, whereas specificity is low [3]. It is true that noninfected collections are most often encountered in this setting, such as serous, lymph, urine, bile or digestive liquids. These collections are usually anechoic. Their observation alone is usually sufficient for diagnosis. The increase in volume of a collection is one criterion for reoperation in postoperative peritonitis [4]. We simplify the approach by adopting the easy tap policy. At the expense of useless taps (but never deleterious if basic rules are respected), septic or hemorrhagic postoperative complications will be promptly detected.

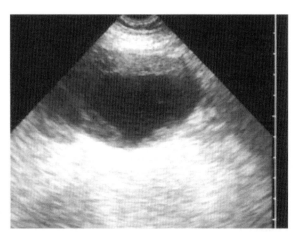

Fig. 23.1. Intra-abdominal abscess in a man operated on for colic ischemia. Transverse scan of the right fossa iliaca. The ultrasound-guided tap was particularly relevant here

The classic subphrenic abscess is rare in our observations.

Acute acalculous cholecystitis is probably a complication particular to the surgical ICU.

Forgotten foreign bodies will easily be detected. A compress gives a large image with a matrix-like pattern and a massive acoustic shadow. A metallic instrument has a strikingly straight shape, with typical posterior artifacts we call S lines.

Hematomas are first anechoic, then rapidly become echo-rich and yield heterogeneous, solid images. They can be observed in the retroperitoneum, the pelvis, and the rectus abdominis muscle.

Postoperative Abdominal Interventional Ultrasound

A simple tap will confirm infected collections. Percutaneous drainage under ultrasound guidance deserves to be subsequently tried. The fluidity helps in choosing the appropriate caliper of the material [5]. This kind of procedure can preclude subsequent surgery, which has higher morbidity and mortality rates. This is the best procedure for some [6], who reserve conventional surgery for complex cases, or when a percutaneous route appears dangerous (bowel obstacles, for instance).

Before inserting a large drain, it can be advantageous to withdraw the maximum amount of pus with a fine needle, which will in certain cases be considered sufficient.

Postoperative Thoracic Ultrasound

Hemothorax, pneumothorax, tamponade, phrenic paralysis, pneumomediastinum, some false aneurysms (see Chap. 19) and sometimes mediastinitis are accessible with ultrasound.

In the postoperative thoracic period, the intensivist must promptly determine if the content of the hemithorax is fluid or air. Ultrasound immediately provides the answer.

A periaortic collection can be detected and even tapped with ultrasound guidance. Sepsis of the prosthesis will thus sometimes be diagnosed. In this severe setting, the current habit is, however, to perform CT, despite its invasiveness.

Here again, appropriate information to the team limits the extent of the dressings.

Thromboembolic Disorders

Lower Extremity Veins

Ultrasound is more laborious in surgical patients than in medical patients, especially trauma patients, as the dressings, surgical devices, pain and postcontusion changes can decrease the potential of ultrasound. Deep venous thrombosis, however, seems more frequent in the surgical ICU, perhaps because local trauma is a major cause for venous thrombosis. It must be remembered that compression ultrasound can be painful, and Doppler may have an interest here.

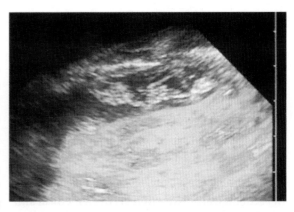

Fig. 23.2. Massive thrombosis of the left internal jugular vein in a patient who underwent venous catheterization. Note that this thrombosis is completely occlusive and extends at least 6 cm in the craniocaudal axis

Upper Extremity Veins

A frequent problem in the emergency setting is the difficulty of inserting a central venous catheter. In surgical ICUs, patients have already been managed. Hypovolemia has been corrected. Therefore, problems in inserting venous lines may not be as critical as in the medical ICU.

In our experience, the frequency of internal jugular venous thrombosis seems extremely high in severely ill surgical ICU patients (Fig. 23.2, and see Figs. 12.6, 12.9, 12.10, 12.13, pp 72–74). Independent factors may explain this, such as the possibly more frequent use of cardiac catheterization in certain surgical ICUs.

References

1. Kox W, Boultbee J (1988) Abdominal ultrasound in intensive care. In: Kox W, Boultbee J, Donaldson R (eds) Imaging and labelling techniques in the critically ill. Springer-Verlag, London, pp 127–135
2. Weill FS (1989) Echographie abdominale du post-opéré. In: Weill FS (ed) L'ultrasonographie en pathologie digestive. Vigot, Paris, pp 536–544
3. Mueller PR, Simeone JF (1983) Intra-abdominal abscesses: diagnostic by sonography and computerized tomography. Radiol Clin North Am 21:425–431
4. Dazza FE (1985) Péritonites graves en réanimation: modalités du traitement chirurgical. In: Réanimation et médecine d'urgence. Expansion Scientifique Française, Paris, pp 271–286
5. Van Sonnenberg E, Mueller PR, Ferrucci JT (1984) Percutaneous drainage of 250 abdominal abscesses and fluid collections. Radiology 151:337–347
6. Pruett TL, Simmons RL (1988) Status of percutaneous catheter drainage of abscesses. Surg Clin North Am 68:89

Ultrasound in Trauma

In the trauma context, ultrasound has a limited place in patients who are lucky enough to arrive alive at a hospital where a CT whole-body examination is readily available. CT in fact answers a majority of questions at the head, thorax and abdominal levels. However, the extreme handiness of a small, autonomous ultrasound device makes it possible to envisage a major role on site. In addition, it is undoubtedly useful to invest time in ultrasound if in the future CT has limited access for reasons of irradiation. All abdominal and thoracic and even cephalic disorders have ultrasound expression.

Thoracic Trauma

On site, ultrasound detects disorders requiring immediate management: hemothorax, pneumothorax, and selective intubation. A tamponade can be found easily as well as aortic rupture provided there is a favorable morphotype. Early signs of lung contusion are available. This is useful since early radiograph misdiagnoses these alveolar-interstitial disorders in 63% of cases [1]. Myocardial contusion can also give signs in two-dimensional ultrasound.

Diaphragmatic Rupture

A diagnosis of diaphragmatic rupture creates a challenge that CT and MRI are far from solving. Ultrasound has no precise place here. Lacking experience, we cannot assess this area. The only comment to be made is that the diaphragm is almost always detectable using ultrasound in critically ill patients (see Figs. 4.9, p 22, 15.5 and 15.7, pp 98 and 17.2 and 17.15, pp 117 and 126).

Abdominal Trauma

In this context, detection of peritoneal effusion is such a basic step that it sums up the role of ultrasound in pre-hospital use [2]. Fluid detected in the peritoneal cavity is usually blood, but urine, bile or digestive fluids can give effusions in trauma patients.

The rupture of a hollow organ gives pneumoperitoneum.

The other findings should be dealt with separately. Analysis of the various parenchymas depends on the patient's morphotype and digestive gas. A parenchymatous contusion (liver, spleen, or kidney) gives a heterogeneous, rather hypoechoic than hyperechoic image (Fig. 24.1). Fracture of a parenchyma can yield a fine hyperechoic line (Fig. 24.2). A pancreas trauma gives the same patterns as acute pancreatitis. A subcapsular hematoma gives a hypoechoic image in a biconvex lens. The diagnosis of vascular pedicle rup-

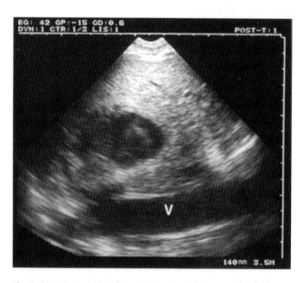

Fig. 24.1. Liver contusion. Heterogeneous ragged image within the liver parenchyma in a patient with abdominal trauma. *V*, inferior vena cava

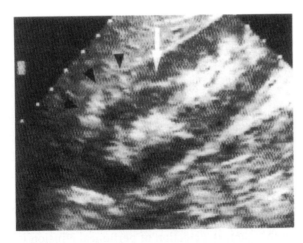

Fig. 24.2. Kidney fracture. The clear line (*white arrow*) indicates a virtual space at the level of the fracture. The *black arrowheads* delineate the hematoma of the renal space

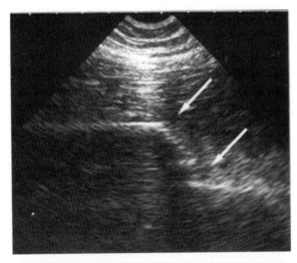

Fig. 24.3. Displaced fracture of the femoral diaphysis. The proximal and distal segments are 20 mm distant, without overriding (*arrows*). Real-time analysis clearly depicts this type of lesion

ture, especially at the kidney, is usually better approached by Doppler and other imaging modalities (CT or angiography).

Cervicocephalic Trauma

The brain is not really accessible to ultrasound, but optic nerve analysis can give information on a possible brain edema. Eyeball integrity can be checked using ultrasound. A solution of cervical vertebra continuity is also accessible to ultrasound from C1 to C7.

On-site checking for this accurate vertebra pile can provide vital information before CT on rachis stability. A traumatic dissection of the carotid artery can be detected using two-dimensional ultrasound alone, but we lack data to confirm this. The hemosinus, cranial dish-pan fracture and many other points will undoubtedly be documented in the future.

Bone and Soft Tissue Trauma

Ultrasound can, if necessary, detect long-bone fractures (Fig. 24.3). Bones have a complex geometry, but at certain levels such as femoral diaphysis, ultrasound can analyze the cortex with accuracy. A minimal solution of continuity can be detected by scanning. Ultrasound makes no pretense of replacing radiography, inasmuch as the probe can be harmful. However, in the sedated patient, this is no longer a problem, and the field of ultrasound is again broadened.

Indeed, a very wide-ranging domain needs to be created, with an investment in bone ultrasound that intensivists may not wish to undertake. On the other hand, it is not excluded that the coming decades will see the emergence of a new type of specialist who will be able to considerably simplify numerous situations where only radiography or CT supplied the answers, and in the radiology department.

Let us imagine a few situations: recognition of a cranial dish-pan fracture, a displacement of the cervical rachis (see Fig. 21.10, p 156), a long bone fracture (femur, tibia, fibula, humerus, radius, cubitus, fingers, etc.), even a rib fracture all give specific ultrasound signs. Multiple cases can be imagined from the most vital (odontoid) to the most functional (scaphoid). For each of these cases, radiography can provide solutions, but we are sure that ultrasound holds surprises in reserve.

With swelling of a limb, ultrasound can settle between hematoma, muscular contusion and venous thromboses.

Whole-Body Exploration: CT or Ultrasound?

Many authors highlight the role of CT in the initial assessment of the polytraumatized patient [3, 4]. CT provides a complete study of the deep organs, the skeleton (especially the cervical spine), a functional study by iodine injection that shows vascular ruptures or parenchymal lesions at the liver, spleen, kidneys, etc. CT is more easily accepted (once the patient is on the table) since ultrasound can be harmful here.

However, CT is reserved for the most stable patients, i.e., the least severely traumatized. Unstable patients are those who will definitely benefit from an immediate on-site ultrasound scanning (see Chap. 25). Let us recall that 20% of thoracic trauma cases do not arrive alive at the hospital

References

1. Schild HH, Strunk H, Weber W, Stoerkel S, Doll G, Hein K, Weitz M (1989) Pulmonary contusion: CT vs plain radiograms. J Computed Assist Tomogr 13:417–420
2. Rose JS, Levitt MA, Porter J et al (2001) Does the presence of ultrasound really affect computed tomographic scan use? A prospective randomized trial of ultrasound in trauma. J Trauma 51:545–550
3. Société de Réanimation de Langue Française (1989) Echographie abdominale en urgence, apports et limites. In: Van Gansbeke D, Matos C, Askenasi R, Braude P, Tack D, Lalmand B, Avni EF (eds) Réanimation et médecine d'urgence. Expansion Scientifique Française, Paris, pp 36–53
4. Société de Réanimation de Langue Française (2000) Stratégie des examens complémentaires dans les traumatismes du thorax. In: Léone M, Chaumoitre K, Ayem ML, Martin C (eds) Actualités en réanimation et urgences 2000. Elsevier, Paris, pp 329–346

Emergency Ultrasound Outside the Intensive Care Unit

The intensive care unit is only the first step for practicing and developing emergency ultrasound.

Ultrasound in the Emergency Room

The development of ultrasound in the emergency room can solve many situations. For the moment, the critically ill patient admitted to the ER will be quickly taken in charge by the intensivist. Respiratory distress, circulatory shock, coma, acute renal failure, drug poisoning, pneumothorax and others are situations where the patient is usually managed directly by the intensivist.

On the other hand, countless situations that do not depend on intensive care medicine and are managed by the emergency physician will be simplified by the use of ultrasound. Pneumonia, renal colic, venous thrombosis, rib fracture, an impressive number of situations can be quickly diagnosed or quickly ruled out. It is to be expected that a rational use of ultrasound in the emergency room can solve the problem of the accumulation of patients at the emergency room, an important part of the public image of the hospital.

The surgeon called at the ER considers ultrasound a beneficial tool that will reinforce her clinical sense. Acute appendicitis [1], intestinal obstruction, pneumoperitoneum are some examples among many others.

Note that ultraportable units are a false solution to a real problem: in the ER, there is enough room for the 1978 technology units such as the ADR-4000.

The place for non-ICU emergency ultrasound will not be limited to the ER alone.

Pre-hospital Ultrasound

In a helicopter or an airplane, room is a true concern, and ultraportable units may be advantageous. The first experiment in emergency extra-hospital ultrasound was, to our knowledge, practiced with the described logistics [2]. Using an ultraportable ultrasound unit, the emergency physician directly answered vital clinical questions on site. The aim of this experiment was to analyze the percentage of clinical questions ultrasound answered. Some items such as pneumothorax, hemothorax, hemopericardium, and acute hypovolemia (inferior vena cava caliper) were investigated and provided the answer to 90.6% of the questions. Therefore and without error, the first pre-hospital ultrasound emergency diagnosis was given in the desert, was for pneumothorax, and was done in January 1996.

Air Medicine

This first experience of pre-hospital ultrasound came in fact from the air [2]. It was performed from a helicopter over Africa. This small helicopter had enough room for our ultraportable ultrasound unit, which in fact fit in a small bag. The local conditions (vibrations, possible interferences) in no way affected the ultrasound examination. In many countries with low-density population, physicians (flying doctors) willingly use the air route, and may feel reinforced by this supplementary tool.

Physician-Attended Ambulances

What was possible in a small helicopter is also possible in an ambulance. Should one be destitute in the full arid desert of Mauritania or highly medicalized on the road in the heart of Paris, one may feel the need for an immediate diagnosis. When the far-reaching possibilities of ultrasound are considered, it is hard to believe that this will not be part of the future. A traumatized patient will be confidently approached, pneumothorax or hemothorax immediately detected, a central venous access promptly inserted in extreme emergency, a dyspneic patient or even a comatose patient properly guided.

A pilot's license, so to speak, will be indispensable, more than ever. On the other hand, the development of sophisticated echocardiography in the ambulance without having first provided the teams with general emergency ultrasound (which includes the main heart emergencies) would be a suboptimal way to exploit ultrasound possibilities.

Pediatric and Neonatal Intensive Care Unit

The use of ultrasound will be highly contributive in pediatric and neonatal ICUs. First, the neonate will benefit from high frequency probes, which means higher diagnostic precision. In fact, the higher the frequency, the better the image. Since the deleterious effects of the ionizing radiation are now established in the child, any noninvasive routine method should be carefully studied [3].

Monitoring the respiratory and cardiac functions, mastering the central veins, the transfontanel route are some of the many points of impact to be investigated. An entire chapter will be devoted to the child in the next edition.

Ultrasound of the World

Ultrasound will be as useful in the wealthy ICUs of the affluent world as in the numerous disadvantaged regions of the world where CT is lacking – or even a simple radiography unit. In this very particular setting, a small unsophisticated device, with a solid padlock, will act as a terminal to make advisedly therapeutic decisions using extremely simplified logistics.

References

1. Puylaert JBCM (1986) Acute appendicitis: ultrasound evaluation using graded compression. Radiology 158:355–360
2. Lichtenstein D, Courret JP (1998) Feasibility of ultrasound in the helicopter. Intensive Care Med 24:1119
3. Brenner DJ, Elliston CD, Hall EJ, Berdon WE (2001) Estimated risks of radiation-induced fatal cancer from pediatric CT. Am J Roentgenol 176:289–296

Interventional Ultrasound

The ICU is a privileged arena for practicing interventional ultrasound. It allows therapeutic management at the bedside of untransportable critically ill patients. It remains, in experienced hands, a safe method [1]. As the patient is, by definition, under high surveillance, early detection of the main complications (hemorrhage or sepsis) is guaranteed. As the procedures are usually done on sedated patients, stress-induced complications [2] are also bypassed.

Interventional ultrasound is indicated for almost every organ. Its positive findings are as valuable as its negative ones.

As regards interventional imaging, CT is preferred by some to ultrasound [3]. Here again, ultrasound offers overwhelming advantages: a bedside procedure, permanent control of the procedure, no irradiation for the patient or for the operator's hands; it is the method of choice for others [4].

Procedures regarding pleural or peritoneal effusions, gallbladder, central veins, etc. were described in the corresponding chapters.

Diagnostic Procedures

The following sites have been routinely investigated at the bedside of our patients: pleura (including patients on mechanical ventilation), pericardium, peritoneum, gallbladder, hepatic abscess, splenic abscess, retroperitoneal hematoma, soft tissue abscess, and mediastinitis through sternal disunion.

Following basic rules and using ultrasound guidance, complications caused by the procedure were absent. Diagnostic relevance was significant, and a retrospective study will confirm this utility.

Therapeutic Procedures

The following percutaneous procedures are routinely used: drainage of purulent pleurisy and pericardial tamponade, aspiration of abdominal abscesses (liver, spleen, pancreas or peritoneum), and percutaneous nephrostomy. Percutaneous cholecystostomy is recommended by some authors.

Patient Management

Patient management can be greatly facilitated by ultrasound:

- Insertion of central venous catheter:
  - Internal jugular
  - Subclavian
  - Femoral (when there is no arterial pulse)
- Insertion of suprapubic catheter
- Right heart catheterization
- Insertion of a Blakemore probe
- Caval filter insertion and percutaneous gastrostomy, although these remain theoretical

Other Diagnostic Procedures

A parenchymal biopsy may provide emergency documentation of tuberculous miliary, hepatic metastases.

Basic Technique for an Ultrasound-Guided Procedure

For the rather large collections (i.e., projection of 4 cm^2 or more), an ultrasound-guided landmark can be established, followed by the puncture. Once the landmark has been determined, the patient must remain strictly in the same position, the ultrasound unit is switched-off, the skin is disinfected and the needle is inserted. This procedure concerns the large majority of pleural or peritoneal effusions. It has the advantage of great simplicity and can be done in a few instants without help.

For smaller targets such as central veins, the procedure is performed under permanent ultrasound guidance. There are probably several protocols. We will describe our technique, carried out by a single operator. We prefer this procedure, since the maneuvers and encounters between the probe and the needle are coordinated by one person alone.

The operator puts on sterile clothes and installs the operative field.

If real asepsis is not fully controlled, interventional ultrasound is impossible. The probe as well as a long part of the cable should be protected. This was a major problem. We were not able to find a system dedicated to these applications on the market. A sterile glove is not a viable solution. The solution came from the combination of a sheath initially dedicated to a video camera and a transparent adhesive dressing (OpSite type), which is not an obstacle for ultrasound beams and closes the opened end of the sheath. This solution is as elegant as it is efficient. It requires less than 2 min to set up.

The procedure can begin. General anesthesia should not exempt from local anesthesia. Betadine proves to be an effective contact product (sterile gel can again be used). The operator holds the probe in one hand, the needle in the other, and proceeds to the tap.

Once the needle is in the target, the probe can be released. It is laid down on the field, because it is sometimes necessary to use it again.

In rare and delicate cases, a second operator will help in aspiring the syringe, whereas the needle and the probe are firmly held by the first operator.

Targeting

Any procedure must be planned: where should the probe and the needle be applied? Which anatomical structures will be pierced? How far will the needle enter? This can be rapidly checked.

The Target

The probe must be applied so as to settle the target comfortably in the sights. Small needle-angle errors are more easily corrected. In other words, the target should not disappear from the screen if small movements are applied to the probe.

The target must sometimes remain motionless (perfect precision is needed in lesions near the diaphragm). One major advantage of our setting is that perfect apnea is extremely easy to obtain: the ventilator can be disconnected a few seconds, which results in the target being strictly motionless. In a spontaneously breathing, tired patient, such apnea would be hard to obtain.

Relation Between Probe and Needle

Servo-control systems exist. The needle is inserted through a device fixed on the probe. We do not use this kind of system but instead use one hand to hold the probe, one hand to insert the needle, and visually direct the needle. This has the following advantages: very simple material, great flexibility, as the operator is free to make slight changes in needle inclination, for instance, and above all, the use of the same device as in the 25 previous chapters. The operator must invest, however, in understanding the position of objects in space. The needle must follow the sole plane of the probe. The needle can be more or less parallel to the probe, more or less far, but the needle must remain in the probe's plane.

Skin-to-Target Distance and Needle Length

The distance that will be covered by the needle to reach the target can be measured (see Fig. 15.10, p 100). In fact, the distance the needle is inserted must be slightly longer than the distance measured on the screen. One explanation is that the probe pulls the soft tissues and brings the skin nearer the target, which the needle does not do. To be precise, the distance read on the screen should be multiplied by a correction factor (of roughly 1.2–1.4) that depends on the patient's adiposity and the pressure that must be exerted on the probe to have an image.

Location of the Point of Needle Insertion and Needle Angulation

With the probe firmly held and the target fully on the screen, the needle penetrates exactly in the section plane of the probe, which is defined by the lateral landmark. Thus, the needle remains visible on the screen for the entire length of its trajectory (see Fig. 12.17, p 77).

The correct angle between the probe long axis and the needle long axis, the correct distance between the probe head and the needle tip can be extremely variable. After making complex calculations, we now use a more intuitive method. It is simplest to first apply the probe is just over the target. The angle between needle and probe long axes is 45°. The distance d between the point of needle insertion and the probe head is equal to the distance between the probe and the target (vertical route). The needle length L must be at least equal to:

$L \geq d\sqrt{2} \times$ correction rate
i.e. roughly $L \geq 2\,d$

If the needle does not go straight toward the target, it should be withdrawn as far as possible and inserted again with a corrected angle. Making angulations without withdrawing the needle could lacerate the soft tissues.

Visualization of the Needle Within the Soft Tissues

A minor problem, as we will see, is that the needle is not always perfectly visible on the screen when it crosses the superficial tissues, which concerns small targets such as a subclavian vein. During venous catheterization, the needle is perfectly visible in two-thirds of cases, but it is more difficult to detect it in the last third. This situation comes up whether the needle is thin or thick, the gain is low or high, the patient is thin or overweight, whether the patient is on corticotherapy, or whether an anesthetizing product has been locally injected.

When there is evidence of a hardly visible needle, our attitude is in fact very simple. The progression of the needle through the soft tissues is followed, in the exact continuation of the plane of the probe. At a precise moment, the tip of the needle is seen when the proximal wall of the vein (in venous procedures) is reached, and when the needle enters the lumen. The problem is solved. Small maneuvers can help meanwhile. The operator can give fine to-and-fro movements to the needle, which can help visualize the needle. It is also possible to hold the needle directly, without it being mounted on a syringe, with the critical condition that there is no risk of gas embolism (see Chap. 12). This maneuver provides fine-tuned control of the procedure.

To sum up, this situation should not be a source of complications. There are indeed many solutions. Some manufacturers produce grainy needles, which can be detected more easily in these circumstances. Some inject small quantities of air in order to locate the end of the needle. This may rapidly render the area impossible to interpret. Color Doppler has been described as an aid [5], as well as the cabled transmission of an electric source at the end of the probe [6, 7]. Calling on CT in this particular situation remains of questionable value. This would result in transportation, greater cost, and irradiation reaching both the patient and the operator.

Penetration of the Target

When a needle crosses a parenchyma, visualizing it is usually easy (see the case of a subclavian vein, Fig. 12.17, or the pericardium, Fig. 20.17). When the target is an encapsulated collection whose deep wall is concave toward the probe, such as a urinary or gallbladder target, echo ghosts can sometimes be generated. These echoes can be unsettling since they mislead the operator, who may risk blindly inserting the needle. A moderate gap between the middle and the end of the needle can be seen with certain probes. Experience is required to distinguish real from artifactual echoes.

The needle always shifts the proximal wall of the vein, the gallbladder, etc. slightly before piercing it. Some recommend inserting the needle roughly. We never like to be rough. If done, however, this procedure requires very careful monitoring of the structures located posterior to the target.

Equipment for Percutaneous Drainage

Whenever possible, we use the simplest material. Experience shows that the majority of emergency procedures require very simple material. For withdrawing pleural effusions, we find it elegant to use a very fine, 16-gauge catheter, which is withdrawn at the end of the procedure. The sophisticated materials used in the radiology department are rarely indicated.

Just one word will be said about these techniques. There are a great number materials on the market. The catheter should be rigid enough to avoid plications during skin and aponeurosis crossing. Its caliper should be adapted to the type of collection. The material used is in fact complex to understand, since the caliper is expressed either in F (for French), where number and caliper increase in parallel, or it is expressed in G (for gauge), and here it is just the opposite: the smaller the caliper the greater the gauge. It would have been simpler to measure all instruments in millimeters. The catheter ends straight or with a pigtail. It can be multiperforated. It is introduced using the trocar or the Seldinger method.

Whenever possible, we use a 16-gauge Cathlon, 60 mm in length, inserted according to the trocar principle. The important point is to try to avoid subcutaneous bayonet-shaped routes, which would end the procedure. Prolongations are welcome since the distal end of the Cathlon will gain in freedom. This material is sufficient for the majority of pleural and peritoneal effusions, and, in thin patients, pericardial and gallbladder procedures.

In other cases where very thick liquid is suspected, a Pleurocath (a thin chest tube) may be preferred. We have also been able to successfully drain hepatic abscesses using an 8-F catheter pigtail with lateral holes (Fig. 26.1). With these materials, it is wise to check that the pieces are perfectly adapted before the procedure. Cathlon, guide, dilatator and catheter are sometimes furnished in kits – not the cheapest solution.

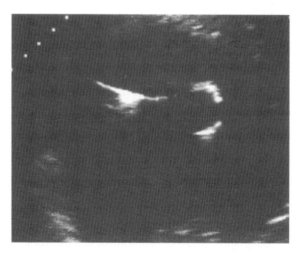

Fig. 26.1. Pigtail catheter inserted within the hepatic abscess shown in Fig. 7.5, p 42

General Precautions Before Any Puncture

All precautions are a part of normal procedure. They include:

1. The nature of the target. Vascular, hydatid, endocrine (pheochromocytoma) masses must be suspected before the puncture makes the diagnosis, in a rather dramatic way. In our experience, the vascular nature of a mass can be suspected if it is possible to detect an echoic flow, with, for instance, whirling movement within a false aneurysm (see Fig. 19.12 and corresponding text, p 137). A slow movement without real flow can result from the passive mixing in a closed collection (plankton sign, see pp 101). In case of doubt, a Doppler study can be done.
2. The right indication. It is not possible to detail this vast domain. Experience plays a large role. Rules can change. Let us recall, however, that in a critically ill patient with pleural or peritoneal effusion for instance, an easy puncture or easy tap policy will always be beneficial. It is important to recognize that a collection is fluid, which makes it possible to use minimally traumatizing equipment.
3. The areas crossed. If the pleura is crossed when an abdominal collection is punctured via the intercostal route, it can be contaminated. At the middle axillary line, the pleura can reach the tenth rib [8]. It should be remembered that ultrasound is an excellent tool for detecting the lower aspect of the pleura.
   Pleural effusion must not be punctured if there is interposition of the lung.
   The internal mammary (thoracic level) or epigastric (abdominal level) vessels should be avoided. Ultrasound can detect precisely the location of these vessels using a 5-MHz probe.
   When the gallbladder is punctured, a transhepatic approach limits the risk of biliary leakage in the peritoneum (the technique is detailed in Chap. 8).
4. The necessary apnea. If a mobile target must be reached in a spontaneously breathing patient, the procedure must be done under apnea, which is difficult to obtain. If the patient has to breathe again, the needle should be released, in order not to lacerate the soft tissues by a firmly held needle [9].
5. Precautions regarding hemostasis. Hemostasis is usually impaired in very critically ill patients. If the basic rules described above are respected, such troubles are never an obstacle to an ultrasound-guided procedure. In 14 years experience, the only side effect induced was the transfusion of two blood units in a patient in whom a postprocedure compression was not performed.

References

1. Nolsoe C, Nielsen L, Torp-Pedersen S, Holm HH (1990) Major complications and deaths due to interventional ultrasonography: a review of 8000 cases. J Clin Ultrasound 18:179–184
2. Barth KH, Matsumoto AH (1991) Patient care in interventional radiology: a perspective. Radiology 178:11–17
3. Dondelinger RF, Kurdziel JC (1993) Drainage percutané des collections abdominales guidé par l'imagerie. In: Actualités en Réanimation et Urgences. Arnette, pp 3–15
4. O'Moore PV, Mueller PR, Simeone JF, Saini S, Butch RJ, Hahn PF, Steiner E, Stark DD, Ferrucci JT Jr (1987) Sonographic guidance in diagnostic and therapeutic interventions in the pleural space. Am J Roentgenol 149:1–5
5. Hamper UM, Savader BL, Sheth S (1991) Improved needle-tip visualization by color Doppler sonography. Am J Roentgenol 156:401–402
6. Winsberg F, Mitty HA, Shapiro RS, Hsu-Chong Y (1991) Use of an acoustic transponder for ultrasound visualization of biopsy needles. Radiology 180:877–878
7. Perella RR, Kimme-Smith C, Tessler FN, Ragavendra N, Grant EG (1992) A new electronically enhanced biopsy system: value in improving needle-tip visibility during sonographically guided interventional procedures. Am J Roentgenol 158:195–198
8. Nichols DM, Cooperberg PL, Golding RH, Burhenne HJ (1984) The safe intercostal approach? Pleural complications in abdominal interventional radiology. Am J Roentgenol 141:1013–1018
9. Duvauferrier R, Carnos C, Delperrier-de Korvin B, Guibert JL (1986) Guidage sous échographie. In: Duvauferrier R, Ramée A, Guibert JL (eds) Radiologie et échographie interventionnelles. Axone, Montpellier, pp. 39–45

CHAPTER 27

Emergency Ultrasound and Antibiotic Therapy

The study of infectious diseases is a basic field of intensive care. We will not insist on the necessity of an early, accurate and noninvasive diagnosis of the infectious concerns that surround the critically ill patient. In septic shock, it has been demonstrated that early and adapted antibiotic therapy is a priority as compared to symptomatic measures (fluid therapy, vasoactive drugs, etc.), and delays in this antibiotic therapy result in an increased death rate [1, 2].

General ultrasound plays a first-line role in this step. It allows a diagnosis of the sepsis site at the bedside and, at the same time, provides an ultrasound-guided sample, which is promptly sent to the laboratory. This authorizes immediate treatment, not probabilistic but adapted.

Many sites are suitable for this two-step approach. We will detail the example of pleural effusion. It is extremely frequent at admission of a patient with acute respiratory failure, but it is rarely recognized on the usual radiographs. It is usually detected on subsequent radiographs or on CT, in a patient already on probabilistic antibiotic therapy. When recognized at admission, pleural effusion is not often taken into account by the intensivist who desires not to cause damage by a hazardous tap. If one considers all patients with pleural effusion, with or without antibiotic therapy, the diagnostic tap identifies a microorganism with an extremely high frequency, 16% in our data. If care is taken to perform this tap at admission, this rather high rate will dramatically increase.

Another eloquent example is pneumonia. Pneumonia is a compact mass swarming with microbes. The usual procedures (plugged telescopic catheter) result in false-positives and false-negatives, and an additional risk of pneumothorax. A transcutaneous tap of certain alveolar consolidations, provided rules are respected, generally withdraws a pure culture of the responsible germ. A randomized study is planned on this very point.

Mediastinitis will be immediately diagnosed if the collection extends lateral to the sternum. It should be remembered that the internal mammary vessels can be avoided using ultrasound location. Bacterial pericarditis can be recognized in the same way.

At the abdominal level, the detection of peritoneal effusion, particularly when small, should be followed by exploratory tap as soon as there is clinical suspicion of infectious process.

A parenchyma abscess of the liver or spleen can be recognized and punctured at admission.

A collection developing in a patient suffering from acute pancreatitis raises the old question of whether it is an abscess or a simple necrosis. In the favorable cases, an ultrasound-guided tap answers this question.

The gallbladder is a classic target in the ICU, but we have seen that the ultrasound data alone are not solid enough for ordering surgery or deciding on abstention. Ultrasound-guided bile aspiration is not a risky procedure when basic rules are respected, although we lack sufficient experience to say whether this procedure is contributive or not.

A septic syndrome occurring around a retroperitoneal hematoma can be a superinfection. An ultrasound-guided tap is possible in some cases, and can avoid the need for CT.

Occasionally, ultrasound detects an abscess of the soft tissues, which will be punctured following the same logic.

Last, still within the idea of searching for a septic site, maxillary sinusitis, endocarditis and several other diagnoses are accessible to two-dimensional ultrasound.

All these applications have several features in common. They can be achieved at the bedside, promptly, with simple equipment and without irradiation. Maximum safety is ensured using the permanent visual control that ultrasound provides. Clearly, the whole body can benefit from the diagnostic and interventional approach of ultrasound.

References

1. Cariou A, Marchal F, Dhainaut JF (2000) Traitement du choc septique: objectifs thérapeutiques. In: Actualités en réanimation et urgences 2000. Elsevier, pp 213–223

2. Natanson C, Danner RL, Reilly JM, Doerfler ML, Hoffman WD, Akin GL, Hosseini JM, Banks SM, Elin RJ, MacVittie TJ et al (1990) Antibiotics versus cardiovascular support in a canine model of human septic shock. Am J Physiol 259:H1440–H1147

Chapter 28

Analytic Study of Frequent and/or Severe Situations

Our small, compact ultrasound device allows for whole-body exploration. How does it work in practice?

Exploration of Septic Shock or Febrile State at Admission

Ultrasound can rapidly give basic information, detecting pleural effusion, pneumonia, peritoneal collection, rupture of hollow organ with pneumoperitoneum, acute cholecystitis, biliary or urinary obstacle, and acute disorders of the liver, spleen, kidney and pancreas. Cardiac vegetation, mediastinal collection or soft tissues are sometimes recognized as the source of sepsis. Nearly all of these findings can benefit from ultrasound-guided diagnostic tap.

Ultrasound of an Abdominal Disorder

Nearly all major painful abdominal syndromes give ultrasound signs. Ultrasound gives more information than plain radiography and is often able to replace CT when the surgery is indicated. A methodic analysis should be carried out of the following structures.

The Wall

A parietal hematoma or abscess can sometimes simulate intra-abdominal emergencies.

The Peritoneum

A search for gut sliding and peritoneal analysis can detect pneumoperitoneum, peritonitis or hemoperitoneum, all disorders usually requiring surgery.

The Bowel

Many items are accessible:

- Peristalsis
- Wall thickening
- Loop caliper
- Liquid contents
- Intraparietal gas
- Intrahepatic gas
- Gastric repletion

They contribute to the following diagnoses:

- Mesenteric infarction
- Occlusion
- Pseudomembranous colitis
- (Theoretical) bullous pneumatosis
- (Theoretical) complicated gastroduodenal ulcer
- Acute gastric dilatation

Other Hollow Organs

Cholecystitis, angiocholitis, obstacle of the upper urinary tract and bladder distension are common diagnoses.

Plain Organs

Liver, spleen and sometimes kidney abscesses are usually rapidly diagnosed. Acute pancreatitis gives signs in the best cases.

Vessels

- Leakage of abdominal aortic aneurysm
- Mesenteric venous thrombosis

Retroperitoneum (Aorta and Kidneys Excluded)

- Retroperitoneal hematoma

Thoracic Disorders with Abdominal Expression

Inferior myocardial infarct, pneumothorax, pleural effusions or pneumonia can sometimes mislead and suggest surgical abdominal emergencies. Each of these diagnoses can be handled by ultrasound.

Exploration of a Thoracic Pain

Pain is assumed to be intense since the patient is managed by the intensivist.

Aortic aneurysm, aortic dissection, pericarditis, myocardial infarct and esophageal rupture give characteristic ultrasound signs, as well as the thoracic disorders seen above (pneumothorax, pleural effusion, pneumonia).

Ultrasound Exploration of Acute Dyspnea

Ultrasound exploration of acute dyspnea is not yet routine. All the skill of the operator is required here, since the examination, performed in a distressed patient, should neither delay nor mislead the treatment. This assumes an on-site ultrasound device: a small one, not too small, adapted to the emergency. Obviously, the operator must be experienced. These points assembled, little time is lost. If no time must be lost, ultrasound examination can be performed instead of the physical and radiographic examinations, possibly saving time. Therefore, regardless of clinical and even radiological data, which are sometimes precious but other times misleading, ultrasound provides objective data that allow the physician to identify the cause of the dyspnea by detecting:

- Pneumothorax
- Acute pulmonary edema
- Cardiogenic or lesional origin of pulmonary edema
- Substantial pleural effusion
- Alveolar consolidation
- Atelectasis
- Pulmonary embolism
- Pericardial tamponade
- Exacerbation of a chronic obstructive pulmonary disease
- Obstacle visible at the cervical trachea
- Acute gastric dilatation
- Acute hypovolemia, with the cause identified at the same time: digestive fluid sequestration, internal hemorrhage
- A »nude« profile sometimes seen in dyspnea accompanying sepsis (with possibly ultrasound-visible site of sepsis) or metabolic acidosis

All in all, although this notion is not routine, ultrasound can provide accurate diagnosis of acute dyspnea in a majority of cases. The use of ultrasound immediately provides the accurate diagnosis in 85% of cases, whereas the traditional approach can only claim accurate diagnosis in 52% of cases [1].

The flow chart we have established uses an exclusively dichotomous design to lead to the accurate diagnosis (Fig. 28.1). Note that our data were obtained without including the right heart status and using only two-dimensional information (limited to contractility) on the left heart.

The Case of Pulmonary Embolism

When pulmonary embolism is suspected, general ultrasound alone in our experience is basic. General ultrasound:

1. Rules out other diagnoses resulting in pain, dyspnea, shock, etc. Pneumothorax, pneumonia, acute pulmonary edema, rib fracture, abdominal disorders (splenic infarction, for instance) or any site of sepsis is ruled out at the first use of the probe.
2. Provides the diagnosis. Many signs are available at the bedside. Pulmonary embolism is certain in the exceptional cases where a thrombus is seen in an ultrasound-accessible right pulmonary artery. Pulmonary embolism is nearly certain when a peripheral, more or less floating thrombus is detected. Suggestive signs are a dilated right ventricle, and above all negative signs such as a normal lung surface, i.e., anterior A lines with lung sliding.
3. Suggests logical management of an extremely frequent case: when there is weak suspicion of nonsevere pulmonary embolism in a patient who is not suffering from acute dyspnea. This is the case of an isolated lower thoracic pain.

This reflects our practice over the last 12 years in which we have not encountered unpleasant surprises. We first check the patency of all accessible venous axes, then check for the presence of a correct margin of respiratory safety, and finally plan simple clinical follow-up, resulting in one of two situations. Either another diagnosis becomes clear (fever appears, positive hemocultures, etc.), or if suspicion remains, then we request pulmonary

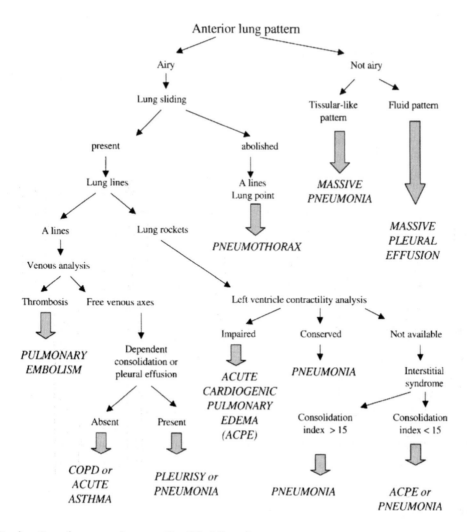

Fig. 28.1. Exploration of an acute dyspnea. Simplified flow chart

scintigraphy or even spiral CT in a well-prepared patient. This outlook can avoid the nocturnal angiography or spiral CT, or worse nocturnal heparin therapy before confirmatory examinations. All these procedures have a mortality and a morbidity rate that is increased by the emergency setting. Our outlook already has one merit: an extreme simplification of the immediate management. Elementary logic indicates that a patient with free venous axes cannot suddenly worsen minor pulmonary embolism. The unpleasant surprises certainly come from massive and unstable thromboses of the large vessels.

Combining General and Cardiac Ultrasound

A stethoscope can be applied indiscriminately at the lung, heart or abdomen. Similarly, the ultrasound probe can be used extensively. The physician gains in synergy and exponentially increases efficiency. This has been demonstrated above for acute dyspnea exploration and will now be explained for cardiac arrest.

One characteristic situation where this synergism is found is anuria. Using four items, this typical situation can be checked in a few instants. First of all, urinary probe permeability must be checked, as a probe obstruction is always possible. Second, lung rockets are sought. Absence of lung rockets indicates that the patient is not in lung overhydration. This indicates that a fluid therapy will not have immediate negative consequences for the lung. A flattened inferior vena cava will indicate hypovolemia. Roughly, a hypercontractile left ventricle provides the same information, whereas an hypocontractile left ventricle suggests a low cardiac output as a cause of anuria.

Ultrasound in Cardiac Arrest

Carrying out an ultrasound examination during resuscitation in cardiac arrest is not yet a reflex.

The information provided in extreme emergency will not be useful to the physician who perfectly and accurately controls the following points. This paragraph is rather devoted to the young intensivist on call who discovers a patient on whom no previous information is available at the moment of action.

The usual maneuvers must not be delayed. Ultrasound will clearly be harmful if it involves any therapeutic delay. Therefore, it is in striving for the objective of gaining time where every second counts that ultrasound can be most advantageous.

The ultrasound device should be moved to the bedside by one member of the staff while the resuscitation is undertaken. Obviously, the more the device is cumbersome and complex, the less it will be used.

When cardiac output is interrupted, the blood is visible in the vessels and the heart chambers. In a few seconds, it takes on an echoic tone (see Fig. 20.24, p 148).

An adiastole caused by tamponade will be promptly recognized and drained.

An asystole caused by cardiac arrest due to massive hypovolemia should be suspected if the chambers have virtual volume.

Can rhythm or conduction problems be detected in this context? Since the answer is not absolutely negative, ultrasound may be highly relevant. Clearly, potential signs that we do not yet know how to recognize may exist, and it is not excluded that a precise ultrasound sign will be found in the future, complementary to the EKG, which is not always readily available. Our observations need to be supported by large studies. In ventricular tachycardia or electromechanical dissociation, observations seem to show barely detectable ventricular contractions. Asystole and ventricular fibrillation seem to yield complete absence of motion. Torsade de pointe seems to give moderate but regular contractions. Many applications should be developed as a priority. For instance, genuine ventricular fibrillations can give asystole on EKG, a potentially valuable indication, but solid data is required for confirmation.

The heart is not the only target involved in this setting.

Tension pneumothorax responsible for cardiac arrest can be instantaneously ruled out. Only 1 s is usually required per lung. Precious time is saved.

When there is massive hypovolemia, the detection of internal hemorrhage (e.g., hemothorax, hemoperitoneum) can be made in a few seconds and authorize massive fluid therapy.

The insertion of a central venous line, if decided, should be successful at the first attempt or not undertaken. This is completely unforeseeable. Three options are possible: ultrasound-guided catheterization with traditional material, or simple checking for a favorable venous caliper, or again immediate insertion of a 60-mm-long catheter in the subclavian or jugular vein (see p 79, Chap. 12). These maneuvers take a few seconds and do not really interrupt cardiac massage. If an internal jugular or subclavian route are chosen, about 10 s is required for an experienced operator. Note that for less experienced operators, the femoral route can be used with ultrasound guidance, as the arterial pulse is no longer here to clinically locate the point of insertion.

If an EES probe must be inserted, ultrasound has a double advantage: venous access first, guiding the probe within the cardiac cavities second.

Once cardiac activity is restored, the same information (pneumothorax, hemothorax, tamponade, venous access, etc.) should be searched for in a calmer atmosphere. Selective intubation will be checked using ultrasound.

The wise reader can ask if this use of ultrasound will decrease mortality and morbidity (e.g., neurological sequela) of cardiac arrest. Substantial proof is rightfully required. The setting (hospital or street), the difference in patients, the great difference in management style from one physician to another (despite the written recommendations), and the emergence of new treatments all create a predictable situation: no room for evidence-based medicine. It is time to believe in ultrasound or not. For want of randomized series, we cannot but rely on anecdotal evidence, however extensive. We also note that, in the year 2001, experienced centers applied very sophisticated but inadequate treatments, whereas sometimes a modest use of ultrasound would have instantaneously provided a diagnosis that would have escaped the best physicians.

Ultrasound and Deciding on Fluid Therapy

Issues on evaluating blood volume have been briefly discussed in Chaps. 13, 17 and 20. We will try to go further here, without excessively simplifying this worldwide debate. It should be noted that much remains to be said, but that a practical alternative is possible, eventually open to criticism, but of immediate use in the emergency setting.

Analysis of the hemodynamic status raises a number of issues. The absence of a gold standard is not the least of these. In the critically ill patient, blood pressure, peripheral edemas, hematocrit, etc. are very unreliable signs. Hemodynamic investigation therefore uses sophisticated techniques. Traditional right heart catheterization gives, it is true, precise and reliable data, but the very meaning of these data are questioned [2]. Hence, the invasive character of the Swan-Ganz catheter generates a questionable risk-benefit ratio [3]. A modern trend is to turn hemodynamics into a noninvasive technique (or semi-invasive) by carrying out transesophageal ultrasound. This approach is potentially very interesting. However, this does not provide a sudden cure for all problems. The logistics is complex (cumbersome, costly equipment and long training for staff), resulting in a still marginal penetration. In addition, the information obtained does not answer all questions [4]. Even if the answer is binary at the end of a transesophageal echocardiography (TEE) (no hypovolemia vs hypovolemia), a doubt frequently remains; everyone has seen flagrant inadequacies of the method. A frequent problem in the ICU is the patient with data (especially wedge pressure) that are not characteristic of a single frank status [5]. Discussions that are complex and impassioned, if not venal, revolve around the respective advantages of invasive vs semi-invasive techniques for assessing the value of a particular parameter [6]. The current struggle that opposes the two techniques seems to have become chronic, whereas the emergence of new approaches such as the PICCO (which measures lung water) or fine analysis of arterial pressure [7] shows to an absurd extreme that the problem is not considered solved.

By integrating unsophisticated cardiac, lung and venous data, we propose an approach that should be compared with more subtle ones. Left ventricular contractility, inferior vena cava caliper and anterior lung surface signal are analyzed. A typical profile of hypovolemia associates small hypercontractile left ventricle, flattened inferior vena cava and complete absence of lung rockets. These data must certainly confirm the clinical impression. However, basic signs such as heart rate sometimes reveal how complex correct interpretation can be; not surprisingly, we often trust the ultrasound information only. Years of practice confirm this approach. Let us insist on a basic point: inappropriate fluid therapy classically risks pulmonary edema. The ultrasound absence of lung rockets does not indicate that such a patient must have fluid therapy. It only tells us that this patient can have fluid therapy (i.e., without risking pulmonary edema). This nuance, although not highly academic, is appreciated in real emergencies, where time lacks for sophisticated answers.

One possible application among others is the hyponatremia seen at the ER. The classic question of dilution or depletion is raised. Lung rockets are highly suggestive of dilution hyponatremia, with a patient in pulmonary subedema, even without a clinical or radiological sign perceptible yet. Inversely, complete absence of lung rockets is highly suggestive in this context of depletion (or at least, absence of hyperhydration).

Exploration of Acute Deglobulization

For acute anemia, ultrasound has rapid access to all possible sites of hemorrhage by detecting effusion that can prove valuable if substantial in a patient with signs of shock.

Hemothorax, hemoperitoneum, sometimes hemopericardium, capsular hematoma (liver, spleen, kidneys), retroperitoneal hematoma, soft tissue collection, and even GI tract hemorrhage with gastric or bowel inundation is quickly recognized. The following step, if needed, is confirming the hemorrhagic nature of the effusion using a safe diagnostic tap.

Conversely, a normal ultrasound scan brings to mind other causes for a drop in hemoglobin (hemodilution, hemolysis, etc.).

Ultrasound in a Patient With Gastrointestinal Tract Hemorrhage

Ultrasound is not mandatory for managing this GI tract hemorrhage. In some cases, a whole-body approach can be useful:

- Early diagnosis of hypovolemic shock (see »Ultrasound and Deciding on Fluid Therapy«).
- Early diagnosis of GI tract hemorrhage, before any exteriorized bleeding (Chap. 6).
- Immediate insertion of a venous line (possibly central) in a hypovolemic patient (Chap. 12).

- Diagnosis of esophageal varices, cirrhosis, detection of indirect signs of gastroduodenal ulcer (Chap. 6).
- Guidance for inserting a Blakemore probe (Chap. 6).
- Early detection of complications stemming from the Blakemore probe: left pleural effusion (Chap. 15), left pneumothorax (Chap. 16).
- Detection of an abdominal aortic aneurysm (Chap. 10), with leakage in the GI tract, a rare finding, but with immediate therapeutic consequences.
- Detection of enolic dilated cardiomyopathy (Chap. 20), a possible association, which can result in a bad adaptation to acute hypovolemia.
- GI tract hemorrhage can be secondary to septic syndrome. A sepsis site can be detected at this occasion (see above).
- Finding hepatic metastases can be fortuitous, since the patient's history is not available in an emergency, and can have consequences on the management (Chap. 7).
- Monitoring gastric content (Chap. 6).

As seen, local conditions govern whether ultrasound can be used.

Contribution of Routine General Ultrasound in a Long-Stay Intensive Care Unit Patient

Many problems can plague a prolonged stay in the ICU. Fever, fall of diuresis, increase in creatinemia, jaundice, poor adaptation to the ventilator, digestive occlusion, edema of lower extremity, edema of upper extremity, low cardiac output, deglobulization, septic shock or multiple organ failure are some of the outward signs.

Ultrasound can be of help in almost all of these situations. It can be negative, thus avoiding more complicated tests (for example, absence of biliary obstacle in case of jaundice). It can be positive, objectifying an abdominal sepsis site, acute acalculous cholecystitis, peritonitis or any other infected collection, urinary obstacle, nosocomial pneumonia, septic pleurisy, lung abscess, pneumothorax with mechanical ventilation, deep venous thrombosis of the lower extremities, lymphangitis and superficial venous thrombosis due to peripheral perfusion, deep venous thrombosis on indwelling catheter, maxillary sinusitis, and more.

Finally, the other causes of fever such as bedsores are superficial and today are not a matter for ultrasound.

A routine exploration can draw up an »ultrasound photograph« which, like a regular physical examination, detects newly emerging alterations.

Difficult Weaning

Ultrasound can detect a number of conditions earlier than radiography:

- Diffuse interstitial syndrome (hydrostatic surcharge or pneumonia)
- Substantial pleural effusion
- Occult pneumothorax
- Voluminous but radio-occult alveolar consolidation, usually located behind the diaphragm
- Phrenic dyskinesis
- Venous thrombosis (of any territory), a source of small but iterative emboli
- Substantial peritoneal effusion, hampering phrenic excursion
- Maxillary sinusitis, a possible source of pneumonia

All these situations can delay weaning.

Pregnancy

We will end by this infrequent but highly awkward situation. The possibility of pregnancy in a critically ill female is raised in Chap. 9. Once it is known that the patient, admitted for instance for lung injury, is pregnant, what is the best course to follow? This is the very time to carefully read the ultrasound device's user's manual. This noninvasive method should now be considered as yielding a decisive answer, and no longer simply a harmless but approximate test requiring confirmation.

The list of the complications that can be directly managed with ultrasound analysis alone is edifying.

- This young patient can develop pneumonia (aspirative or nosocomial), which is recognized, quantified and watched over under therapy. Repeated ill-defined radiographs are eliminated.
- Pleural effusion can be diagnosed and directly drained, with no need for diagnostic radiographs or even CT, follow-up X-ray after thoracentesis, or imaging procedures needed to detect complications such as pneumothorax due to blind thoracenteses.

- If intubation is necessary, selective insertion of the tracheal tube is detected or ruled out using ultrasound.
- Iatrogenic pneumothorax can be detected, drained and followed up without the traditional procedures (repeated radiographs or even CT).
- Abdominal complications such as cholecystitis, hollow organ perforation, peritonitis etc. bring the patient to the surgeon directly, thus avoiding the usually uninformative plain abdominal radiographs as well as highly irradiating CT.
- A subclavian catheter is inserted and monitored using ultrasound, thus avoiding follow-up X-rays and the numerous radiographs for the various possible complications.
- The correct position of a gastric probe can be checked using ultrasound.
- Venous thromboses, acute dyspnea due to pulmonary embolism directly benefit from heparin, reducing the need for venography or spiral CT.
- Maxillary sinusitis will no longer need CT.

To sum up, if radiological procedures must be forgotten, they can be, provided the patient benefits from the ultrasound assistance alone. The particular case of pregnancy clearly demonstrates that ultrasound can open the way to a genuine visually based medicine.

References

1. Lichtenstein D, Mezière G (2003) Ultrasound diagnosis of an acute dyspnea. Critical Care 7 [Suppl] 2:S93
2. Jardin F (1997) PEEP, tricuspid regurgitation and cardiac output. Intensive Care Med 23:806–807
3. Connors AF Jr, Speroff T, Dawson NV, Thomas C, Harrell FE Jr, Wagner D, Desbiens N, Goldman L, Wu AW, Califf RM, Fulkerson WJ Jr, Vidaillet H, Broste S, Bellamy P, Lynn J, Knaus WA (1996) The effectiveness of right heart catheterization in the initial care of critically ill patients. J Am Med Assoc 276:889–897
4. Boldt J (2000) Volume therapy in the intensive care patient – We are still confused, but… Intensive Care Med 26:1181–1192
5. Michard F, Teboul JL (2000) Using heart-lung interactions to assess fluid responsiveness during mechanical ventilation. Crit Care 4:282–289
6. Magder S (1998) More respect for the CVP (editorial). Intensive Care Med 24:651–653
7. Perel A (1998) Assessing fluid responsiveness by the systolic pressure variation in mechanically ventilated patients. Systolic pressure variation as a guide to fluid therapy in patients with sepsis-induced hypotension. Anesthesiology 89:1309–1310

CHAPTER 29

Learning and Logistics of Emergency Ultrasound

The introduction of emergency general ultrasound in an intensive care unit should not be improvised. Usually, the current logistics combines a radiologist and a complete, cumbersome ultrasound device in the radiology department. The ultrasound device is provided with wheels, but using these wheels is quite another matter. This set-up is effective when the radiologist is skilled in emergency ultrasound signs, and is physically present day and night, and when the patient can be transported without harm to the radiology department.

In an indeterminate number of institutions, even in high-income countries, the radiologist is little accustomed to emergency ultrasound, is reluctant to let the equipment leave the radiology department, or is absent outside of normal working hours. In this precise configuration, a more active role for the intensivist can be envisaged. A suitable ultrasound unit, suitable training and suitable checking of standards could then be combined.

The Ultrasound Unit

Chapter 2 described the ultrasound unit. The acquisition of a device in the ICU assumes a financial investment. Occasionally the radiology department gets rid of obsolete units and leaves them to whoever wants them: these »old« machines can save lives. Their acquisition is a temporary but sometimes extremely interesting solution.

Training

Intensivists can be trained in emergency ultrasound. The training must progressively become part of their day-to-day practice. Ultrasound mastery has certainly a beginning but no end. This author continues to learn every day. However, from the beginning, basic steps can be acquired one after the other. To begin with, training can be limited to a single application, for instance lung sliding in the search for pneumothorax. Once accustomed, the intensivist knows that the device can be used every time this precise question is raised. Once fully familiarized, the intensivist will go on to another application, and so on for an indeterminate period. To give a rough estimate, personalized training including one 30-min session every week will cover the 12 basic applications in 18 months [1]. The time required to master a single application can be extremely short.

The training of the intensivist in emergency ultrasound assumes a global reflection. This training can be acquired by reading books devoted to emergency ultrasound. Classic training among colleagues in the same ICU is probably the best, but not many will be trained per year. Seminars may accelerate this process. In fact, integrating ultrasound use into university medical studies would be the most efficient way to prepare future intensivists.

The Pilot's License

Untamed ultrasound is expanding more and more. This means that the intensivist comes up to an ultrasound device, switches it on, carries out the examination and uses conclusions for immediate management. These conclusions may be compared with other diagnostic tools (time permitting) or with a follow-up ultrasound examination performed by authorized personnel. This practice is difficult to control and can give eminently variable results depending on the operator's experience and conscientiousness. Usually performed in the anonymity of nighttime on-call duty, this practice has undoubtedly saved many critical situations throughout the world.

Controlled access to this type of ultrasound use will be hard to apply, since deontology rules should be adapted. The deontology code indicates that no one should go beyond one's abilities, but in cases of extreme emergency, all possible means must be put to service. We strongly believe that becoming an intensivist implies a very particular motivation. The same forces that pushed toward this discipline with admittedly few rewards will likewise motivate to combine self-control and conscientiousness. It is hoped that the appropriation of this life-saving method will give the user a feeling of humility, and not the opposite. The wise reader will beware of the danger of tarnishing the method [2, 3]. Let us wager that the number of situations saved with ultrasound will exceed the number of cases where the ultrasound device should not have been switched on.

Meanwhile, the future organization of a university certificate will allow the intensivist to practice this discipline with the approval of the medical community, but it is as yet unknown exactly what official place ultrasound holds in extreme emergency situations.

References

1. Lichtenstein D, Mezière G (1998) Apprentissage de l'échographie générale d'urgence par le réanimateur. Réan Urg 7 [Suppl] 1:108
2. Filly RA (1988) Ultrasound: the stethoscope of the future, alas. Radiology 167:400
3. Weiss PH, Zuber M, Jenzer HR, Ritz R (1990) Echocardiography in emergency medicine: tool or toy? Schweiz Rundschau Med Praxis 47:1469–1472

Ultrasound, a Tool for the Clinical Examination

Ultrasound cannot and must not replace the physical examination. It is not conceivable to practice ultrasound before having clinically examined the patient. However, in emergency medicine, one absolute aim is to proceed quickly and accurately. We can therefore meditate on ultrasound's capability to extend, not to say surpass, the physical examination in certain instances.

The physical examination has critical advantages (no cost, innocuousness, etc.) but also some limitations, all the more worrying as we are examining a critically ill patient. Pulmonary edema without crackles, hemoperitoneum without provoked pain, venous thrombosis without clinical signs, urinary obstacle without pain, or, more simply, all the difficulties arising from an examination performed in obese or ventilated, sedated patients are situations where the physical examination can show itself to be insufficient. In addition, the information obtained from years of training is immediately confirmed – or refuted – when the intensivist holds the ultrasound probe.

Let us consider the ultrasound device as if it was a clinical tool, a kind of stethoscope.

Half of the work will be done if one considers that an examination performed at the bedside is a clinical examination, in the etymological sense. The other half will be achieved if one looks into the meaning of the word »stethoscope«, which comes from the Greek and was created by a French physician at the beginning of the nineteenth century. This instrument, which has symbolized medicine for nearly 200 years, strictly means »to observe throughout the chest wall«.

Considering ultrasound an extension of the physical examination is becoming widespread. Let us make a brief overview of the services ultrasound can offer when considered this way.

The Abdominal Level

A peritoneal effusion is promptly detected, long before dullness of the flanks appears.

Prompt identification of diffuse air artifacts replaces the clinical search for tympanism.

Visualization of peristalsis makes the search for air–fluid sounds unnecessary – a sign that may be of low sensitivity.

The often difficult search for a hepatomegaly is replaced by the direct ultrasound detection of an enlarged liver, which can also reveal its origin (tumor, abscess, right heart failure, etc.).

An area that is sensitive to palpation (or echopalpation) will reveal the cause: parenchyma abscess, cholecystitis.

The search for pain from the shaking of the liver no longer has a raison d'être if a liver abscess has been identified, and the patient will be grateful to us!

Going farther, we could say that the free hand of the operator can also evaluate abdomen suppleness or, on the other hand, parietal contraction.

The Thoracic Level

The basic elements of lung examination, i.e., inspection, palpation, percussion, auscultation, are reinforced if ultrasound detects pneumothorax, pleural effusion or alveolar consolidation. As regards interstitial syndrome, only ultrasound can recognize it, as there is no clinical equivalent.

A heart analysis informs immediately on the pulse and contractility. This may rejuvenate the search for muffling of heart sounds or galloping rhythm. A vegetation may be detected whether or not there is heart murmur. Regardless of whether there is pericardial rubbing (precisely the main feature of substantial effusions), pericardial effusion, its tolerance, and sometimes its origin can be recognized at the same time.

Infinite examples can be cited. Detection of a cardiac liver and of jugular turgescence are redundant with the existence of right chambers dilatation, provided they are not compressed by a pericardial tamponade.

The diagnosis of dehydration can be clinically delicate. It is reinforced by the detection of collapsed venous trunks (inferior vena cava) or heart chambers and a dry lung surface, without interstitial changes.

Certain physical signs such as the increase in precordial dullness belong to the past since ultrasound has entered the emergency setting.

At the thoracoabdominal junction, several combinations can be imagined: a painful right hypochondrium indicates an acute cardiac liver; moving the probe then reveals enlarged right chambers; a shift of the probe at the venous level (e.g., iliofemoral) then detects the venous thrombosis that was responsible for the previous disorders.

The Peripheral Level

A rapid scan along the lower and upper venous axes easily rules out the threat of thrombosis.

The behavior of the femoral artery, when compressed by the probe against the bone, can give another view on arterial pressure. When arterial pressure is normal, the compression does not affect the cross-section. Progressively, the lumen collapses, with systolic expansion despite the probe pressure. At an even lower stage, the artery collapses without resistance.

Occult parietal emphysema can give early ultrasound signs.

Serendipitous Applications

An important advantage of ultrasound (which can, like any device, break down) is that it allows the clinicians to improve their accuracy in the physical examination. It is indeed possible to assess one's clinical skill in real-time. For example, pleuritic murmur can be compared with ultrasound pleural effusion. This could be repeated with a variety of clinical signs.

Comparing chest X-ray and ultrasound can also provide the same critical reading of the chest radiography (assuming that ultrasound is a gold standard).

And the Clinical Examination?

All the examples seen above are but a few of the countless situations where ultrasound performs better than the physical examination. Should we therefore mistrust our hands, eyes and ears? In other words, should we dispense with the clinical examination? Does opposing physical examination and bedside ultrasound make any sense? In the extreme emergency or if overburdened, many items of the physical examination will be redundant and therefore waste time. In these precise situations, we do not hesitate to use ultrasound first. In calmer situations, one must absolutely proceed as usual. However, we must admit frankly that when we do not have our ultrasound unit with us, we feel extremely blind.

The truth may be that we see patients very early in an emergency situation, and this can be a source of great disparity between the signs we learned at school and what we see in the ER or ICU. Ultrasound is accused of being highly operator-dependent. This is probably true, but the physical examination may be even more operator-dependent. Physical examination can be considered a complex and uncertain field. Diagnoses such as early bladder distension or pleural effusion can be recognized by well-trained, intelligent hands, after a long training period. Yet these diagnoses are reached much more rapidly using ultrasound. This critical point has not been sufficiently documented.

Several physical signs will obviously never be replaced by ultrasound, particularly inspection (habitus, skin, etc.) and neurological examination. Indeed, where is the harm in placing a mechanical probe[1] over the tibia in order to explore deep sensitivity, thus leaving the cumbersome tuning fork in the attic?

In addition, the physical examination remains an important psychological step. This direct contact between the physician and the highly stressed patient should unconditionally be preserved. Ultrasound is an opportunity for the radiologist to get even closer to the patient.

We will close this chapter with a thought to our elders. The physical examination was their only diagnostic tool, and they knew (at least the most famous among them) better than us how to exploit its numerous subtleties and secrets.

---

[1] This is no longer possible with the modern ultrasound probes, which do not vibrate.

Chapter 31

Concluding Remarks

Our object in writing this book was to depict all the ways in which a simple ultrasound approach can assist the intensive care physician. Indeed, the title could have been »The 1001 Reasons to Perform General Ultrasound in the Critically Ill Patient«.

Ultrasound is increasingly attracting interest in the field of emergency medicine. However, it has to earn its place with respect to CT and MRI, which provide easy-to-read images and will remain indispensable for some indications. First, one cannot just compare the accuracy of ultrasound with these two heavyweights of modern imaging: one must also consider the risks, the expected benefits and the constraints for the patient. Referral for CT or MRI requires reflection, whereas ultrasound can be ordered without hesitation. Second, ultrasound neatly resolves the paradox that the most severely ill patients are those who might benefit most from CT or MRI but often cannot reach the examination suite. Ultrasound is a bedside procedure and in trained hands yields respectable results.

In our setting, even more so than elsewhere, rapid and accurate decisions are crucial. A judicious ultrasound scan will reinforce the physical examination and outclass radiography and, in most cases, CT. Ultrasound is not just a rough test carried out to decide which allegedly more sophisticated investigation is most appropriate. On the contrary, prompt therapeutic decisions can be taken according to ultrasound findings alone. This is of great import in situations where, previously, one had to rely on clinical experience and basic tools such as a stethoscope, perhaps supplemented by an ill-defined radiograph. To the well-known advantages of ultrasound (low cost, etc.) can be added bedside use and an increasing spectrum of indications, e.g., examination of the lungs.

In the first French edition of this book, published in 1992, we wrote that the place of ultrasound in the ICU was modest. In the intervening years the situation has progressively improved. One can now note the emergence of another paradox: »ultrasound in the ICU« is often understood as meaning »cardiac ultrasound in the ICU«. An increasing number of cumbersome specialized devices can be observed, while the lungs, the veins and other organs and tissues still have little access to the wide-ranging applications permitted by lighter ultrasound equipment. If one is prepared to do without Doppler – a remarkable tool but one which can, as we have seen, backfire against the patient – a simple, unsophisticated device provides a valuable whole-body approach, heart included.

Clearly, the results yielded by ultrasound depend on the skill of the operator. However, let us return to the time when, for instance, auscultation was not part of clinical routine. The situation did not change overnight, but nowadays one can hardly imagine any physician calling in a specialist in auscultation to detect rales and then write a report on which subsequent treatment could be based. Ultrasound is nothing more than a stethoscope, slightly heavier than Laënnec's device and powered by electricity. Plainly some years will pass before ultrasound becomes part of the armory of every physician. However, one can observe that institutions which tentatively begin to apply ultrasound soon integrate it into their daily routine and never go back.

Ultrasound is progressively spreading over the medical landscape. First cardiologists and gynecologists, then some gastroenterologists, dermatologists, and surgeons, e.g. urologists – not forgetting veterinarians, who were by no means slow to appreciate ultrasound's potential. All of these specialties have adopted ultrasound as a daily tool. Ultrasound is now beginning to be used by general practitioners, and some voices – including our own! – speak in favor of its introduction into medical studies. It certainly merits use where it has the most potential for benefit, namely in the critically ill.

The ICU of the 21st century thus has to have a suitable structure, i.e., full-time presence of an operator or, better still, progressive training of one or more members of the intensive care team. This concept will not only discharge our ethical obligations but also, above all, lend a new dimension to our duties. With permanent whole-body scanning at our disposal, the patient will be rendered »transparent« directly after admission. Ultrasound will become our daily tool, since it allows nothing but »visual-based medicine«.

Ultrasound has been plagued by widespread misconceptions that have cast a long shadow over its development – and have certainly not helped to save lives. One simple example is the mistaken belief that air is an enemy of ultrasound, whereas in fact the very opposite is true. The existence of a whole lung semiology can be considered a chance for ultrasound to become a genuine stethoscope. A salutary trend considers ultrasound as tomorrow's stethoscope. Let us just note here that ultrasound is *today's* stethoscope, if we recall the etymology of this word coined by Laënnec in 1819: a means of looking (*scopein*, to observe) through the lung (*stethos*, the chest wall).

Glossary

Anechoic

Free of echo. The tone is black by convention.

Artifact

Artificial image created by the physical principles of propagation of the ultrasound beams. The shape is always geometrical with precise symmetric axes. Artifacts do not correspond to real anatomical structures.

Bat Sign

In the initial and basic step of any lung ultrasound, the bat sign identifies in a longitudinal view the upper and lower ribs (the wings) and, deeper, the pleural line (the back of the bat). This step makes it possible to correctly locate the pulmonary structures in any conditions.

Bed Level (at)

When the probe explores the lateral chest wall in a supine patient and cannot explore more posterior (without moving the patient) because of the bed, the probe is said to be applied at bed level (or FDL). If pleural effusion is visible at bed level, this means that this effusion has substantial volume.

Comet-tail Artifact

This term designates a repetition artifact that is hyperechoic and roughly vertical. It can arise *or not* from the pleural line. It can be short *or not* or spread up without fading.

Consolidation With Dynamic Air Bronchogram

Alveolar consolidation within which hyperechoic punctiform particles (indicating the air bronchograms) have a centrifuge inspiratory movement.

Consolidation With Static Air Bronchogram

Alveolar consolidation within which hyperechoic punctiform particles (indicating the air bronchograms) have no visible movement.

Culminating (Sign, Point)

This term refers to the sky-earth axis and indicates something near the sky.

Dark Lung (Ultrasound Dark Lung)

Situation where a diffusely hypoechoic pattern is recorded at the chest wall, with no static or dynamic element that can affirm a solid or fluid predominance. The radiograph usually shows a white lung.

Dependent (Sign, Point)

This term refers to the sky-earth axis and indicates something near the earth.

Echoic

In principle, a tone with the same echostructure as a reference structure (classically, the liver). Usually, »echoic« designates a structure rather »hyperechoic«, i.e., near a white tone.

Gain

Setting the device to provide a well-balanced reference image. The upper parts of the screen can be lightened or darkened (near gain), as can the lower parts (far gain). Experience alone can conclude that the gain is correctly set.

Hyperechoic

Tone located between the reference pattern (classically the liver) and what is called the white tone.

Hypoechoic

Tone located between the reference pattern and a black, anechoic tone.

Interpleural Variation

See »sinusoid.«

Isoechoic

Tone equal to a reference structure (classically, the liver).

Jellyfish Sign

Visualization of particular dynamics of the inferior pulmonary strip within a substantial pleural effusion. In rhythm with respiration and heart beats, this is reminiscent of a jellyfish.

Lung Lines

A Lines

Hyperechoic, roughly horizontal lines, arising at regular intervals from the pleural line. A lines are opposed to B lines, and the term »A lines« can be used to designate O lines.

A1, A2... Lines

Number of A lines arising from the pleural line. The term »A+ lines« means that at least one A line has been detected.

B Lines

This designates a kind of comet-tail artifact that is precisely defined as arising from the pleural line *and* spreading out without fading to the edge of the screen and erasing A lines.

b Line

This term indicates that only one B line is visible in a view.

B7 Lines

This term designates lung rockets whose elements are about 7 mm apart. Three to four comet tails are thus visible in an intercostal space.

B3 or B+ Lines

This term designates lung rockets whose elements are about 3 mm apart (B3), or even contiguous (B+). Seven to ten B lines fit in an intercostal space.

C Lines

Real image (see this term), poorly echoic, curving, on a centimeter scale, arising from the pleural line.

E Lines

E for emphysema. Comet-tail artifacts, long and without fading, arising from superficial layers located above the pleural line, with the particularity of being aligned, indicating a thin air layer formed within two parietal layers. This pattern should never be confused with B lines.

I Lines

Short comet-tail artifacts arising from the pleural line (vanishing after 1–3 cm).

O Lines

O for non-A non-B. Absence of any artifact, either horizontal or vertical, arising from the pleural line.

W Lines

Comet-tail artifacts, long, without fading, arising from superficial layers located above the pleural line, not aligned. These indicate anarchically organized air bubbles in soft tissues (see E lines).

Z Lines

Z for the last letter of the alphabet. Comet-tail artifacts arising from the pleural line, not erasing A lines and quickly vanishing, as opposed to B lines. A Z line should never be confused with a B line.

Lung Point

Sudden and fleeting appearance, generally on inspiration, of a lung sign with lung sliding and/or lung rockets and/or alteration of A lines, at a precise area of the chest wall where abolished lung sliding and exclusive A lines were previously observed.

Lung Pulse

Visualization at the pleural line of vibrations in rhythm with the heart rate.

Lung Rockets

Lung rockets designate several B lines visible in a single view.

Lung Sliding

Dynamics – a sort of to-and-fro twinkling – visible at the precise level of the pleural line.

Lateralization Maneuver

Maneuver consisting of placing the arm of the supine patient at the contralateral shoulder. Several centimeters of the posterior aspect of the lung are thus accessible and can be explored using ultrasound, probe pointing toward the sky. A lateralization maneuver pleural effusion is a pleural effusion that was not visible at bed level (see this term) and is visible only using this maneuver.

Out-of-plane (Effect)

An image that leaves the plane of the ultrasound beam can give a false impression of dynamics (a pseudo-dynamic pattern). This effect must be distinguished from the true dynamics.

Plankton Sign

Numerous punctiform echoic images within an anechoic or echo-poor collection. These images have slow, whirling dynamics, as in weightlessness.

Pleural Line

Echoic line located between two ribs, slightly deeper (0.5 cm), in a longitudinal view of an intercostal space. It represents the interface between parietal tissues and thoracic air.

Pleuro-consolidation

Detection in one view of a pleural effusion associated with an alveolar consolidation (a frequent association).

Posterior Reinforcement

Echoic pattern of an area located behind a fluid structure.

Posterior Shadow

Completely anechoic image, with an artifactual shape and located behind a bony structure.

Real Image

A real image is, as opposed to an artifactual image, shaped with anatomical rather than geometric lines and patterns.

Real-time

Two-dimensional mode. Acquisition of dynamic images in two dimensions. A posteriori, a video tape recording alone can reproduce the dynamic features.

Rockets

See »Lung rockets«.

Seashore Sign

Time-motion pattern of a normal lung. The parietal layers are motionless and generate horizontal lines (reminiscent of still waves) at the upper part of the screen. The image above and from the pleural line generates a granular pattern (reminiscent of sand) since it reflects lung sliding that propagates to the end the screen.

Sinusogram

Ultrasound visualization of the walls of the maxillary sinus.

Sinusoid Sign

Curve acquired in time-motion at the level of a pleural effusion. The superficial limit (the parietal pleura) is motionless, whereas the deep limit (the visceral pleura) displays an inspiratory centrifuge excursion. One can again speak of interpleural variation.

Splanchnogram

Direct visualization of an abdominal organ when the probe is applied in a supine patient, which means that no free gas (pneumoperitoneum) collects at the abdominal wall.

Stratosphere Sign

Time-motion pattern composed of horizontal lines in an intercostal view. This pattern is reminiscent of a bar code, but a more striking image is a flying fortresses squadron in the stratosphere, a pattern characteristic of pneumothorax.

Time-motion (TM)

Analysis of dynamics passing along a precise line. A posteriori, the reading of the image alone detects the observed dynamics. Time-motion is opposed to two-dimensional observations.

Two-dimensional

A two-dimensional image provides a view in two dimensions, as opposed to a time-motion acquisition (see this term). Also see »Real-time.«

Ultrasound-aided Procedure

A procedure is ultrasound aided when done after ultrasound location, as opposed to a procedure carried out with permanent ultrasound guidance.

Subject Index

A

Abdominal aorta 19, 62
Abscess
   Hepatic 42
   Lung 124
   Soft tissues 157
   Splenic 66
Acoustic enhancement 6
Acoustic shadow 6
Acoustic window 5
Acute adrenal failure 68
Acute cardiogenic or lesional pulmonary edema 122, 178
Acute dyspnea 132, 178
Acute pleural symphysis 109, 124
Acute renal failure 55
Adiastole 145, 180
Adrenal necrosis 68
Adult respiratory distress syndrome 123, 144
Aerobilia 45
Air artifacts, classification 131
Air bronchogram
   dynamic 102, 117
   static 118, 123
Air-fluid level
   abscess 125
   digestive 38
   hydropneumothorax 111
Air medicine 168
Alveolar consolidation 116
Alveolar filling 116
Alveolar recruitment 123
Alveolar-interstitial syndrome 119, 132, 178, 182, 186
   Etiology 126
Ambulance 9, 168
Anechoic 5
Aneurysm
   abdominal aorta 62, 177
   splenic artery 67
   thoracic aorta 134
Antibiotic therapy 175
Appendicitis 168
ARDS 123, 144
Artifacts 5, 106, 116, 131
   classification 131

Asepsis 16
Aspiration (pneumonia) 126
Asthma 108, 144, 178
Atelectasis 79, 118, 123, 178

B

Bat sign 105, 113
Bed level 97
Bile leakage 52
Biliary tract 43
Black ultrasound lung 102, 119
Bladder 22, 58
Bladder distension 58
Blakemore-Linton tube 35, 137, 182
Bone fracture 166
Bowel 34, 36
Brain edema 152, 166
Breakdown (CT, ultrasound) 132, 187
Budd-Chiari 44
Bullous emphysema 108

C

Caliper of the inferior vena cava 83, 146, 180
Candidosis (ocular) 152
Cardiac arrest 112, 148, 180
Cardiac asthma 122
Cardiac gallbladder 50
Cardiac liver 41
Cardiac output 139
Carotid artery 155
Cart 11
Catheterization
   cardiac (Swan-Ganz) 140, 147, 164, 180
   internal jugular 76
   subclavian 76
Cellulitis 157
Central venous pressure 83, 146, 168, 187
Cervical rachis (fracture) 155, 166
Chest pain 178
Chest tomodensitometry 129, 130, 182
Chest trauma 165

Cholecystectomy space 530
Cholecystitis
  acute acalculous 46, 163, 177
  bacterial 51
  gangrenous 51
  lithiasic 53
  and pregnancy 183
Cholecystostomy (percutaneous) 53
Cholestasis 43
Chronic obstructive pulmonary disease 122, 178
Chronic right heart failure 144, 178
Chronic subacute cholecystitis 47
Chest radiography 116, 129
  after catheter insertion 79
Clinical examination 186
Colic ischemia 37
Colic necrosis 37
Colon 34, 37
Coma 153
Comet-tail artifact 6, 106, 113, 119, 131, 143, 158, 178, 181
COPD 122, 178
Coupling gel 11

D

Desert (ultrasound in the) 168
Diaphragmatic cupola 23, 97, 126
Dish-pan fracture 166
Dissection
  aortic 62, 148, 178
  carotid artery 155, 166
Doppler 10, 38, 44, 52, 61, 64, 71, 75, 79, 85, 87, 89, 91, 93, 137, 139, 144, 149, 154, 164, 166, 172, 173, 178, 181, 188
  in acute acalculous cholecystitis 52
  in acute dyspnea 178
  basics 10, 89, 188
  brain 154
  false aneurysm 137, 173
  heart 139, 144, 149
  hepatic veins 44
  inferior vena cava 85
  interventional ultrasound 172
  kidney 61, 166
  lower extremity veins 87
  mesenteric infarction 38
  pancreatitis 64
  trauma 164, 166
  upper extremity veins 71, 75, 79, 91, 93
  volemia 188
Dressing (policy) 163
Duodenum 33
Dynamic air bronchogram 102, 117
Dynamic noise filter 3, 107
Dyspnea 132, 178
Dyspnea (acute) and flow chart 178

E

Easy puncture (policy) 29, 173
Echocardiography Doppler 139, 149, 188
Echoic 5
Echoscopy 5
Ectopic pregnancy 60
Electromechanical dissociation 180
Emergency room 168
Emphysema (bulla) 108
Endocarditis 148, 186, 175, 177, 186
Endovascular ultrasound 145
Epigastric vessels 31, 173
Esophageal rupture 136, 178
Esophageal varices 35, 182
Esophagus
  abdominal 33, 35
  thoracic 136
Ethical obligations 189
Exudate 100
Eyeball 152, 166

F

False aneurysm 64, 137
Feasibility of ultrasound 14
Feeding tube 36
Fever in the ICU 177
Fibrosis (lung) 109
Filter (caval inferior) 83
Flight sign 72
Floating thrombosis 73, 79, 90
Fluid therapy 123, 180
Flying doctor 168
Foreign body 163
Fulminant hepatitis 45

G

Gain 4
Gallbladder 21, 46, 175, 177, 182
Gas embolism 76, 147
Gas tamponade 146
Gastric tube 36
Gastric distension (acute) 36, 177
Gastroduodenal ulcer 36, 181
Gastrostomy 36
Gel 11
General ultrasound of the heart 139
Ghost echo 71, 172
GI tract 33
Gut sliding 31

H

Harmlessness (of Doppler) 11
Heart 139
Helical CT 129, 130, 182
Helicopter 9, 168
Hematoma 157
   retroperitoneal 63, 175, 177
   subcapsular 165
   uterine apoplexy 60
Hemopericardium 165, 168, 181
Hemoperitoneum 29, 60, 165, 168, 177, 181
Hemorrhage
   digestive 36, 39, 181
   internal 181
   retina 152
Hemorrhagic shock 29, 168, 181
Hemosinus 150, 166
Hemothorax 101, 165, 181
Hepatisation (lung) 117
Hepatogram 31
Hepatomegaly 41
Hydatic cyst 43, 173
Hypercalcemia 155
Hyperechoic 5
Hypertension (severe arterial) 68
Hyperthermia 159
Hypoechoic 5
Hyponatremia (dilution versus depletion) 182
Hypovolemic shock 83, 146, 180, 187

I

Infectious concerns 175
Innocuity (of Doppler) 11
Interlobular septa thickening 119
Interstitial syndrome 106, 119, 131, 143, 178, 181, 186
Interventional ultrasound
   bacteriological sample 175
   Blakemore probe 35
   caval filter 83
   deep central veins 76
   gallbladder 52, 53
   gastrostomy 36
   lung 127
   mediastinitis 136
   pericardium 146
   peritoneum 31
   pleural effusion 102
   pneumothorax 112
   postoperative collections 163
   retroperitoneum 63
   spleen 67
   Swan-Ganz catheter 147
   technique 170–174
   tracheostomy 155
   urinary tract 61
   veins 76
Intracardiac thrombosis 147
Intracranial hypertension 152, 166
Irradiation 113, 130, 165, 169
Isoechoic 5

J

Jejunum 34
Jet ventilation 108

K

Kidney 21, 55

L

Learning curve 184
Left ventricular failure 143
Lesional edema (pulmonary) 122, 178
Lesional pulmonary edema 122, 178
Lines (ultrasound artifacts)
   A lines 105, 109, 125
   b lines 120
   B lines 106, 119, 125
   B3 lines 119, 125
   B7 lines 119, 125
   C lines 118, 126
   E lines 113, 158
   O lines 107
   W lines 113, 158
   Z lines 106, 114, 119
Lithiasis (urinary) 56
Liver 20, 41
Lung 116
Lung compliance 123
Lung contusion 165
Lung expansion 108, 123
Lung infarction 126
Lung point 110
Lung pulse 108, 123
Lung rockets 106, 119, 131, 178
Lung segmentation 96
Lung sliding 107, 178, 182
Lung ultrasound 105, 116
Lung-wall interface 105
Lung water 122, 181
Lymph node enlargment 68, 72

M

Malignant hyperthermia 159
Maxillary sinusitis 150, 183
Mediastinitis 135, 175, 177
Mediastinum 134
Melena 39
Mesenteric infarction (ischemia, necrosis) 37, 177
Metabolic acidosis 178
Miliary 43, 66
Morrison's pouch 27
Myasthenia 137
Myocardial infarction 148, 178

N

Neonatal ICU 169
Nephrostomy (percutaneous) 61

O

Occlusion (GI tract) 38, 177
One lung intubation 123, 165, 183
Optic nerve 152, 166
Out-of-plane effect 118
Overdistension (lung) 123
Oxygen diprotonate 11

P

Pancreas 22, 64
Pancreatic necrosis 64
Pancreatitis (acute) 64, 177
Parietal emphysema 113, 158
Parotiditis 154
Pediatric ICU 169
Pericardial drainage 146
Pericardial effusion 145
Pericarditis 145, 170, 178, 186
Peristalsis 34, 37
Peritoneal effusion 27, 60, 165, 177, 182
Peritoneum 19, 27
Peritonitis 30, 177
Pheochromocytoma 68, 173
Phrenic paralysis 126, 182
Physical examination 186
Plankton sign 29, 101
Pleura 96
Pleural effusion 96, 177, 178, 182
Pleural line 105
Pleural symphysis 109, 124
Pleurisy (purulent) 101
Plugged telescopic catheter 127

Pneumomediastinum 137
Pneumonia 116, 126, 144, 168, 177, 178
Pneumoperitoneum 31, 165, 168, 177
Pneumothorax 105, 112, 146, 165, 168, 178, 180, 182
    and cardiac arrest 180
    and gas tamponade 146
    occult 111
    posterior 113
    radioccult 111
    tension pneumothorax 105, 112, 180
    under mechanical ventilation 112
Polycystic disease 56
Portal gas 37, 41
Portal hypertension 41, 67
Postoperative collection 163
Pre-hospital ultrasound 168
Pregnancy 60, 131, 182
Principle precaution 11, 75, 131, 182
Probe frequency 10
Prostate 58
Prosthesis (sepsis) 164
Pseudomembranous colitis 38, 177
Pulmonary artery 137
Pulmonary edema 116, 122, 143, 178
Pulmonary embolism 125, 137, 144, 178, 183, 187
    and upper extremity thromboses 74
    and calf thrombosis 92
Pulmonary infarction 126
Pyelonephritis (acute) 55
Pyonephritis 55
Pyonephrosis 56

R

Radiography 79, 116, 129
Renal cyst 55
Renal pelvis dilatation 56
Repetition echoes 6
Retroperitoneum 63
Reverberation echoes 6
Rhabdomyolysis 55, 159
Rib fracture 166
Right heart failure 144
Rockets (lung) 106, 119, 131, 178

S

SAMU 168
Seashore sign 107
Selective intubation 123, 165, 183
Septic shock 175, 177
Severe arterial hypertension 68
Severe asthma 108, 144, 178
Sinusogram 150

Sinusoid sign 99
Soft tissues 157
Spasm (venous) 78
Spatial resolution 52, 132, 152
Splanchnogram 31
Spleen 22, 66
Splenic infarction 66
Splenic rupture 66
Splenomegaly 66
Static air bronchogram 118, 123
Stages of investigation (lung) 97
Stethoscope 186, 189
Stomach 33, 35, 177
Stress 170
Suprapubic catheterization 61
Surgical ICU 163
Swan-Ganz catheter 139, 147, 181
Swirl sign 38, 111, 125
Symphysis (acute pleural) 109, 124
Systemic resistances 63

T

Tamponade 145, 178, 180
Thai dragoon 58
Thoracic aorta 134, 165
Thoracic CT 130, 166
Thoracic pain 178
Thoracic trauma 165
Thrombolysis 93, 126
Thyroid 155
Torsade de pointe 180
Tracheal obstruction 155
Tracheal stenosis 155
Tracheomalacia 155
Tracheostomy (percutaneous) 155
Transcranial Doppler 154
Transesophageal echocardiography 139, 149, 180, 181
Transfontanel route 169
Transudate 29, 100
Trauma 165
  abdomen 165
  cervical spine 166
  chest 165
  head 166

U

Ulcer (gastric) 36, 181
Ultrasound-assisted, ultrasound-guided, ultrasound-enlightened puncture (catheterization, tap...)
  See Interventional ultrasound
Ultrasound at the emergency room 168
Ultrasound of the World 169

Undernutrition 158
Untamed ultrasound 184
Urinary tract 55
Urinary probe 58
Uterus 60

V

Varices (esophagal) 35, 182
Vein
  cava inferior 19, 82, 168, 146, 181
  cava superior 79
  hepatic 44
  internal jugular 70
  lower extremity 87, 164
  subclavian 75
  superior mesenteric 38
  upper extremity 70, 164
Veins, central venous system 70, 164
Venography 93
Venous compression (technique) 88
Venous catheterization (ultrasound-assisted) 76
Venous thrombosis 71, 75, 83, 89, 164, 168, 182, 183
  calf 91
  femoral 89
  floating 73, 79, 90
  iliac 90
  inferior vena cava 83
  jugular internal 71, 164
  mesenteric 38, 177
  occlusive 72
  septic, superinfected 74
  subclavian 75
Ventricular fibrillation 180
Ventricular tachycardia 180
Venturi effect 85
Visual medicine 189
Volemia 83, 146, 168, 187

W

Weaning (difficult) 182
Wedge pressure 85, 122, 180
Whole-body ultrasound 166

X

X-rays 79, 116, 129, 130, 166, 182

Z

Zoom 3

Printing and Binding: Arti Grafiche Nidasio, Assago (Milan), Italy
Printed in July 2007